AF553688

Chemistry for Agriculture and Ecology

Chemistry for Agriculture and Ecology

M. SATAKE
Faculty of Engineering
Fukui University
Fukui 910, Japan

Y. MIDO
Dept. of Chemistry
Kobe University
Kobe 657, Japan

Consulting Editors

S.A. IQBAL
Dept. of Chemistry
Saifia P.G. College of
Science and Education
Bhopal, INDIA

M.S. SETHI
A.R.S.D. College
University of Delhi
New Delhi-INDIA

1995

Discovery Publishing House
New Delhi (INDIA)

First Published 1995

ISBN 81-7141-266-1

Published by :
Discovery Publishing House
4594/9, Darya Ganj
New Delhi 110 002

Printed in India

Laser Typesetting by : **Computer Codes**, Delhi-110 009
Printed at : **Arora Offset Press**, Laxmi Nagar, Delhi-110 092

Preface

The teaching of Chemistry at the introductory stage becomes each day a more challenging task as the subject matter becomes more diverse and more complex. These challenges have evoked a series of responses – the present set of introductory chemistry monographs is one such. The teaching of chemistry recognises a number of problems that confront those who select text books. In order to overcome these problems, this volume "**Chemistry for Agriclutre and Ecology**" – one of about fifty in the Chemistry Monograph Series – is introduced. Each volume is independent of the others deals with one of Chemistry topics and constitutes a complete entity. Each volume is more comprehensive than can be possible in a single volume text. It is intended to provide a range of topics to cover most undergraduate and chemistry main courses of study. These volumes can be used to enrich the more conventional courses of study.

Suggestions for improvement are welcome and shall be gratefully acknowledged.

Authors

Contents

PART I
GENERAL CHEMISTRY

CHAPTER 1

The Nature of Matter

WHAT IS MATTER ?

Chemistry is the study of the properties of the matter that makes up the earth and the rest of the universe. The chemist is particularly concerned with the ways in which different substances react together, and with the way in which some substances are made up from other simpler ones. When beginning a study of chemistry we need to consider the different types of matter.

Matter is anything that occupies space. The amount of matter is its *mass*. Matter is made up of different *substances*. All parts of a substance are chemically the same. Examples of pure substances are diamonds, common salt and water.

THE STATES OF MATTER

A substance can be either a solid, a liquid or a gas. These are known as **the three states of matter.**

A **solid** has a definite shape with distinct boundaries. Its volume is hardly affected by changes in temperature and pressure.

A **liquid** takes up the shape of any vessel in which it is placed, but like a solid it has a definite volume and is hardly affected by changes in temperature and pressure.

GASES

A **gas** has neither a definite shape nor volume. It assumes the shape of any vessel in which it is placed and, moreover, spreads itself evenly throughout any containing vessel. Thus if a gas is put into a large flask the gas will, in a very short considerably affected by changes in temperature and pressure.

The differing properties of solids, liquids and gases are explained by assuming that all matter is composed of minute particles. In *solids* these particles are in definite positions. They vibrate in these positions but do not move from one part of the solid to another. The state that a substance is in depends on the energy that the substance contains. In the solid state a substance is in depends on the energy that the substance contains. In the solid state substances have the minimum energy content. As a solid is heated, the energy of the solid is increased and the particles vibrate more vigorously, until at a certain temperature, which is always the same for the same substance, the solid changes to a liquid, i.e., it melts. For the solid this temperature is called the *melting point.* If the source of heat is removed the liquid will cool and at the same temperature as that at which the solid melted, the liquid will solidify. For the liquid this temperature is called the *freezing point.*

LIQUIDS

In *liquids* the particles move about freely and do not merely vibrate about a fixed position. The particles do, however, attract one another, and these attractive forces give liquids a definite volume. All the particles in a liquid do not move at the same speed; some move faster than others. Moreover, the particles are continually colliding and exchanging energy and hence their speeds are continually changing. Near the surface of the liquid some particles that happen to be moving rapidly will fly out of the liquid and escape. In this way the liquid *evaporates.* If the temperature of a liquid is increased the particles move more rapidly, and so the rate of evaporation increases. For any substance at a certain temperature, and at a certain pressure, the addition of heat produces no further rise in temperature. At this temperature all of the energy being put into the substance is converting the liquid to a gas, i.e., the liquid is said to vapourize or boil. This temperature, which is markedly affected by pressure, is called the *boiling point* of the liquid. If the heat is removed the gas will revert to a liquid. For a gas this process is called *condensation.* Gases that are at so low a temperature that they can be condensed by pressure alone are called *vapours.*

In *gases* the particles are much farther apart from one another than in liquids or solids, and consequently the attractive forces between particles are much less important. Each gas particle moves more or less independently of all other gas particles around it. This is why gases have no definite shape or volume, and it also accounts for the remarkable property of *diffusion* of gasses. Thus if some hydrogen sulphide is prepared in one part of a laboratory, its smell is very rapidly detected all over the room. The properties of gases and discussed further in chapter 5.

Occasionally when a solid is heated it becomes a gas without passing through the liquid state. This process is called *sublimation.* It occurs when solid carbon dioxide or iodine are heated. Carbon dioxide or iodine vapour on cooling condense directly to the solid state. Water vapour goes directly to the solid state when snow flakes and hoar frost are formed, in the upper atmosphere and on the earth's surface respectively. The most common form of sulphur, a fine yellow powder known as "Flowers of sulphur" is also produced by the condensation of the vapour direct to the solid on a cold surface. The relationships between the states of matter are summarized in Fig. 1.1

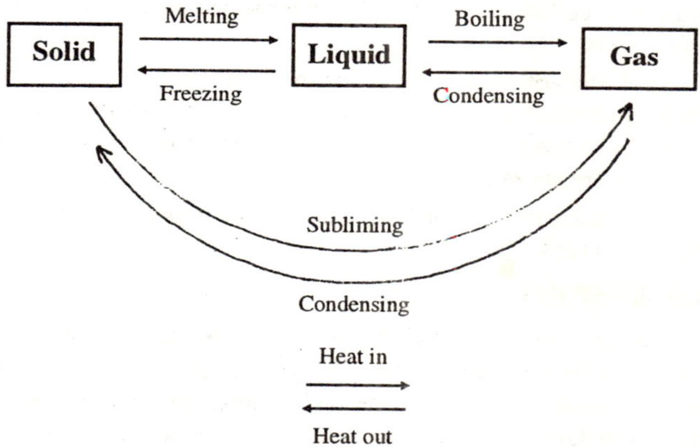

Fig 1.1. The three states of matter.

SOLUTIONS

Some substances, when they are mixed with water, seem to disappear. For example, if a little common salt is mixed water the most careful microscopic examination of the resulting water-like liquid fails to reveal any trace of the salt. Yet testing the liquid gives immediate proof of its presence. It is apparent that the salt and water have formed a mixture which is the same throughout, i.e., *homogeneous*, and the mixture will remain homogeneous even if it is left for ever. Such mixtures are called *solutions*, and the salt is said to have dissolved in the water. The dissolved substance, here the salt, is called the *solute*, and the substance that has dissolved it, here the water, is called the *solvent*.

Water is the most common solvent, mainly because it is so widespread on the earth, but there are many others. If you get a stain on your clothes you might try to remove it with a "stain-remover", or if the clothes are very dirty you may take them to be "dry-cleaned". The stain remover and the liquid used by the dry cleaner will be trichloroethene or some similar organic solvent.

There are many other solvents. Polystyrene, used for making toy model kits, is soluble in propanone (acetone). Cellulose ethanoate (acetate), a substance made from wood pulp, is also soluble in propanone or a mixture of propanone and pentyl ethanoate (acetate). The artificial fibre, rayon, is made by squirting a solution of cellulose ethanoate in propanone through a small hole into a warm atmosphere, when the propanone evaporates and leaves a long fibre behind. Nail varnish is a solution of cellulose ethanoate in propanone, as are ''balsa cement'' and some paints.

Solution of iodine are often in an alcohol solvent. Perfumes also contain pure alcohol as a solvent. Turpentine, or its substitute, white spirit, is a useful solvent for paint; it is often used for diluting paint, and for removing fresh paint stains.

There are many more solvents—you can probably think of others in addition to those give here.

SUSPENSIONS

Many substances do not dissolve in a particular solvent, and so they are said to be *insoluble* in the solvent. Water, for example, has no effect on glass, sand, wood and many other substances. However, if some fine chalk powder, which is insoluble in water, is shaken with it a milky mixture is formed. Such a mixture, where particles are spread throughout a liquid without dissolving in it, is called a *suspension.* These particles will settle out if the mixture is left long enough. Some medicines are suspensions; hence the instruction to ''shake the bottle''. Other common examples of suspensions are muddly water and most paints.

FILTRATION

If a mixture of water and chalk is poured onto a filter paper, the water passes through the filter paper and the chalk particles are left behind. Filter paper is unglazed paper, similar to blotting paper. It contains many small holes, which, although large enough to allow the water to pass through, retain the much larger particles of chalk. This process is called *filtration*; the liquid that has passed through the paper is called a *filtrate* and the solid left behind on the filter paper a *residue*.

Filtration is a much used process, and there are various substitutes for the filter paper depending on the size of the particles to be removed. The most widespread application is in the treatment of our water supplies, where layers of sand and gravel are used to remove suspended particles. In the engine of a car the oil passes through a filter to remove solid particles that if left in the oil would cause damage, and for the some reason dirt particles in the air and petrol are removed by filters before they pass into the engine.

Dust particles are filtered from the waste gases of power stations and other industrial plants to stop this dirt being expelled into the air where it would be a major source of atmospheric pollution. For these industrial uses, where the volume of air to be filtered is very large, a filter paper method would be much too slow. A special type of filter is used called an electrostatic precipitator (Fig. 1.2.)

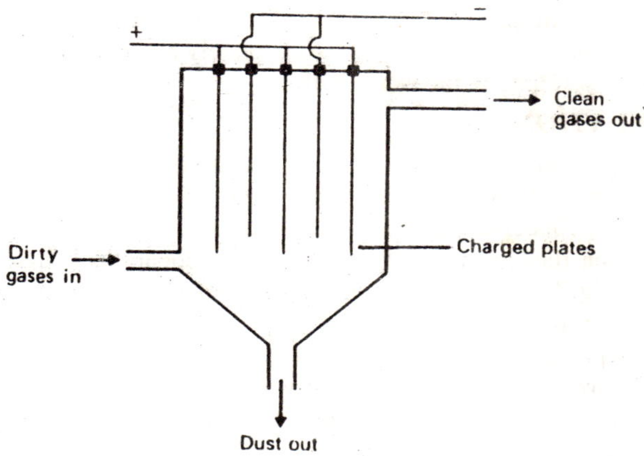

Fig. 1.2. An electrostatic precipitator.

In this the dust particles are electrically charged and they then become attached to metal plates of opposite charge. After a time the electric current through the plates is switched off, and the dust then falls to the bottom of the precipitator when it can be removed.

EVAPORATION

Salt dissolves in water to give a solution. The salt cannot be separated from the water by filtration; they both pass straight through the filter paper. To obtain the salt the solution must be heated, when the water escapes as steam leaving the involatile salt. A porcelain dish called an evaporating dish is often used. If the solution is heated until all the water has been vapourized the process is called *evaporation to dryness*. When doing this care must be taken towards the end; the solution becomes very concentrated and will spit out of the dish as steam is formed beneath semi-solid parts of the mixture. To minimize spitting gentle heating must be used.

As the process of evaporation removes the water, it concentrates the salt. If a solution that has been concentrated is allowed to cool, crystals of the salt may form. This process is called *crystallization*.

When an inflammable solvent, such as ether or alcohol, is being evaporated, the solution must not be heated with a naked flame, but instead a hot water-bath or a steam-bath is used.

A beaker containing water at a high enough temperature to cause evaporation can be used as the steam-bath. Even with a steam-bath great care must be taken when evaporating liquids whose vapours are inflammable and heavier than air, such as ether or carbon disulphide, as these vapours can fall around the water-bath and become ignited. For such liquids the only safe procedure is to completely remove the flame before putting the solution on the water-bath.

DISTILLATION

Evaporation gives only the involatile salt from the salt solution; the water has boiled away into the atmosphere. To obtain the water, it is necessary to condense the steam by cooling. A Liebig condenser, fitted to a distillation flask in normally used, and the apparatus is assembled as shown in Fig. 1.3. A small piece of broken pottery in the flask enables the solution to boil smoothly.

This process is called *distillation*, and the water obtained in this way is *distilled water*. Distilled water is very pure water and is usually obtained by the distillation of tap water. It is used when dissolved solids in the water are undesirable, such as in experiments in the chemistry laboratory and in car batteries.

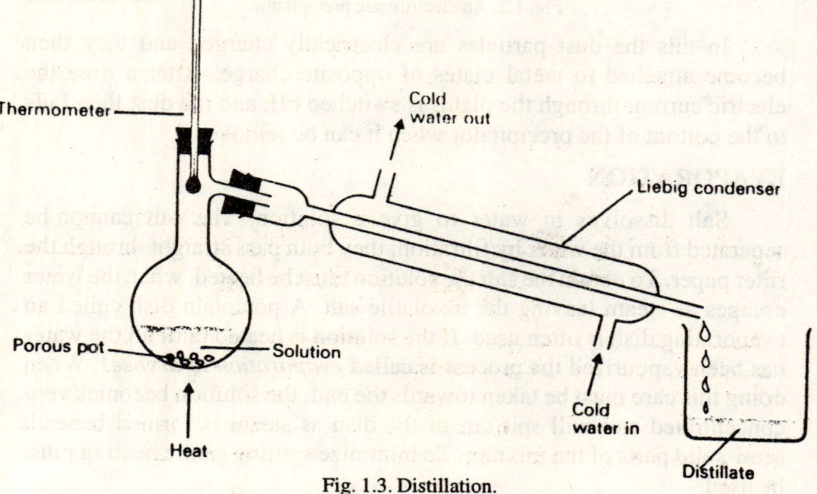

Fig. 1.3. Distillation.

The process of distillation has been used for several thousands of years, mainly to concentrate the alcohol in wine by making spirits. Also, over the past centuries distillation was an important process for the chemists predecessors—the alchemists. The alchemist, in his search for methods of turning other metals into gold a developed the very simple, all glass piece of distillation apparatus called the retort. The retort has become one of the symbols of the alchemist. Paintings of alchemists, such as that by J. Stradanus in the Palazzo Vecchio in Florence abound with retorts and distillation equipment.

Retorts are not used much nowadays, but they are useful when distilling a liquid that attacks metal, rubber and other materials. For example, retorts are sometimes used in the school laboratory for preparing nitric acid and bromine, both of which are highly corrosive liquids. Modern all-glass apparatus with ground joints is, however, just as a good and much more convenient to assemble and to dismantle and clean.

FRACTIONAL DISTILLATION

By distillation it is also possible to separate a mixture of two or more liquids, provided that they have different boiling points. Ethanol boils at a lower temperature than water and when a mixture is distilled the vapour will contain as comparatively high proportion of the lower-boiling ethanol. This vapour condenses to give a distillate richer in ethanol, and the residue in the flask in richer in water. A complete separation is not effected, but if some of the distillate is redistilled a liquid still richer in ethanol is obtained. This is exactly what is done during the making of malt whisky.

Often, however, it is necessary to obtain each of the liquids that make up the mixture in as pure a form as possible. To achieve this a *fractionating column* is used. There are many different types; one of the simplest consists of a long tube packed with small glass beads. This tube is fitted to the neck of the flask containing the liquid to be distilled (Fig. 1.4.) When a mixture of vapours passes up the column those of the less volatile liquid condense on the glass beads more readily than those of the more volatile liquid. Liquid that has condensed on the column is evaporated by the hot vapours, and the more volatile liquid evaporates most readily. As a result of the many evaporations and condensations the vapour moving up the column becomes progressively richer in the more volatile liquid. At the top of the column the vapour will be mainly the more volatile liquid, and this is then collected in the receiver.

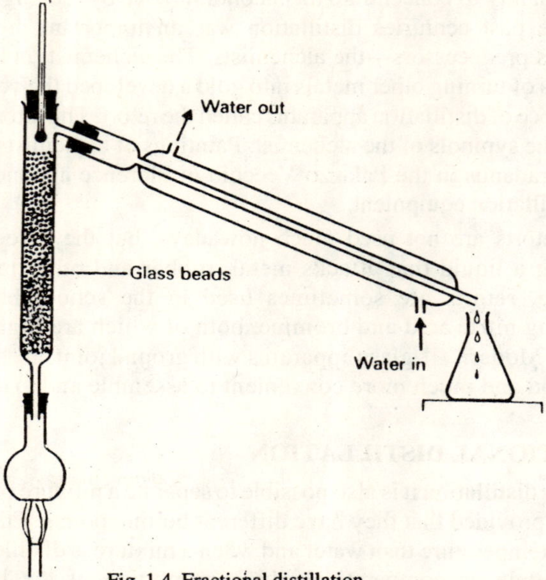

Fig. 1.4. Fractional distillation.

Fractional distillation, as the process is called, is used widely in the chemical industry. Petroleum for cars, aviation spirit for aeroplanes, diesel oil for lorries, paraffin for heaters, oil for lubrication, asphalt for road-surfacing and many other organic compounds are all produced from the oil obtained out of the earth. Of the processes involved in separating the complex mixtures that make up this Crude Oil, the process of fractional distillation is the most important. For these industrial uses the fractionating columns are very large.

If liquid air is fractionally distilled the lower boiling nitrogen is more volatile than the oxygen and collects at the top of the fractionating column; thus these two major components of air can be separated. Wit very efficient fractionating columns the other minor components of air (i.e. the noble gases) can also be obtained in a pure form from the mixture.

SUBLIMATION

When a solid is heated, is becomes a vapour without passing through the liquid state. The vapour will also condense back to the solid without becoming a liquid. This process can be used to purify a mixture of compounds in which one sublimes, but the other does not. For example, if a mixture of ammonium chloride and sodium chloride is heated in a evaporating dish only the ammonium chloride sublimes. The vapour can be cooled by using an inverted filter funnel (Fig 1.4). Solid ammonium chloride collects on the cool sides of the funnel, the sodium chloride remains in the dish.

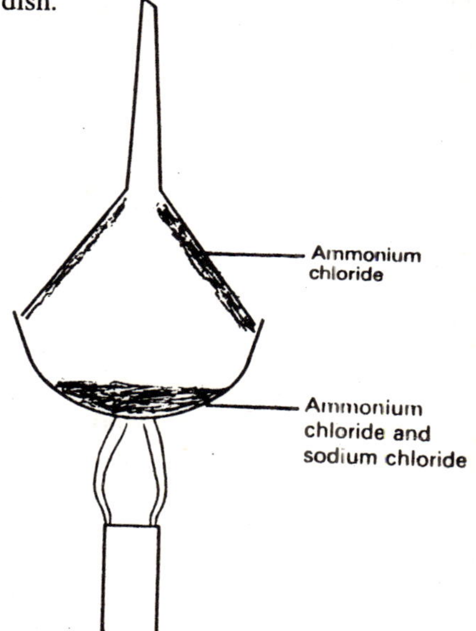

Fig. 1.5. Sublimation of ammonium chloride

Iodine and naphthalene (an organic compound) are two other examples of substances that sublime.

PAPER CHROMATOGRAPHY

This technique is used to separate two, or more, substances that are both soluble in the same solvent. A mixture of dyes can be separated in this way; for example the dyes in a sample of ink (black ink is very good), or screened methyl orange, which is a mixture of two dyes (methyl orange and xylene cyanol (FF)). One method is to cut a filter paper as shown in Fig. 1.6, place a drop of the dye mixture near the centre of the paper and allow the spot to dry. The paper is then placed flat over the top of an evaporating dish with the wick (*A*) dipping into a solvent and covered with a clock glass. Water alone can be used as the solvent but a mixture of water, ethanol, butanol and ammonia is better.

The solvent moves up the wick and then outwards on the filter paper. As it does so it carries the components of the dye with it to different distances from the centre of the paper.

The name *chromatography* comes from the Greek for colour (*chroma*) because the method was first used for mixtures of coloured compounds. Chromatography is now used very widely, and by choosing the right equipment and technique it is possible to separate mixtures of very closely related substances, such as the amino-acids produced when a protein is hydrolysed. Sometimes the substances being separated are colourless, and the chromatogram must then by "developed" to make the different "spots" on the paper visible. An example that requires development is given in the experiment at the end of this section.

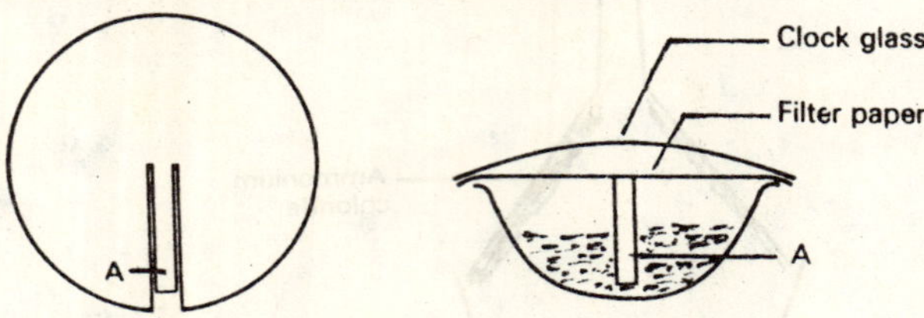

Fig. 1.6. Paper chromatography

The technique of paper chromatography works because the different substances in the mixture show different relative solubilities in two solvents. Within the filter paper is a lot of water; much of this is combined with the cellulose of the paper. As the solvent containing the mixture of solutes passes along the filter paper, the different solutes are soluble to different extents in the water of the paper; they are "partitioned" between the two solvents. The dyes that are most soluble in the water are held back more than the dyes that are less soluble and hence a separation is achieved. It is found that the distance travelled by each solute divided by the distance travelled by the solvent is a constant for a particular solvent and this ratio is called the R, value. The R, value will enable a chemist to say exactly what the substance is.

A chromatography experiment—The separation of iron, cobalt, nickel and copper ions.

In this experiment a more useful technique that the radial method described above, is used. The solvent climbs up the sheet of filter paper from a small volume is the bottom of a beaker (1.7a). The beaker is kept covered so that there is solvent vapour surrounding the chromatogram. Also, the final so that there is solvent vapour surrounding the chromatogram. Also, the final chromatogram must be developed so that the different spots are clearly visible.

Put into the bottom of a 600 cm^3 beaker a depth of 0.5 cm of solvent consisting of 90 parts acetone. 5 parts concentrated hydrochloric acid and 5 parts water (all by volume). On a piece of filter paper, or chromatography paper (21 cm × 11 cm), put a line about 1 cm from the bottom and along this line place small spots of solutions of iron (II) ammonium sulphate (30%), cobalt (II) sulphate (10%), nickel (II) sulphate (10%) and copper (II) sulphate (10%). To obtain small spots use a melting point tube that has been drawn in a Bunsen flame to a fine capillary. Dip the capillary into the solution and then apply the tip to the paper. Put the spots at least 2 cm apart. Allow the spots to dry, roll the paper into a cylinder, fix it together with paper clips and put the paper in the beaker containing the solvent. (Fig. 1.7). Put a clock glass on the beaker as a cover.

Allow the solvent to climb up the paper until the front edge of the solvent is near the top of the paper; this should take about 15 minutes. Remove the paper from the beaker, remove the paper clips, carefully mark on the paper the position of the front edge of the solvent. and allow the paper to dry.

Place the paper in a large covered beaker in which is a small beaker containing concentrated ammonia for about 2 minutes (Fig. 1.7b). Remove it from the ammonia and in a fume cupboard spray lightly with an atomiser spray containing a solution of 0.5 g of dithiooxamide (rubeanic acid) in 100 cm^3 ethanol or industrial spirit, when the spots as

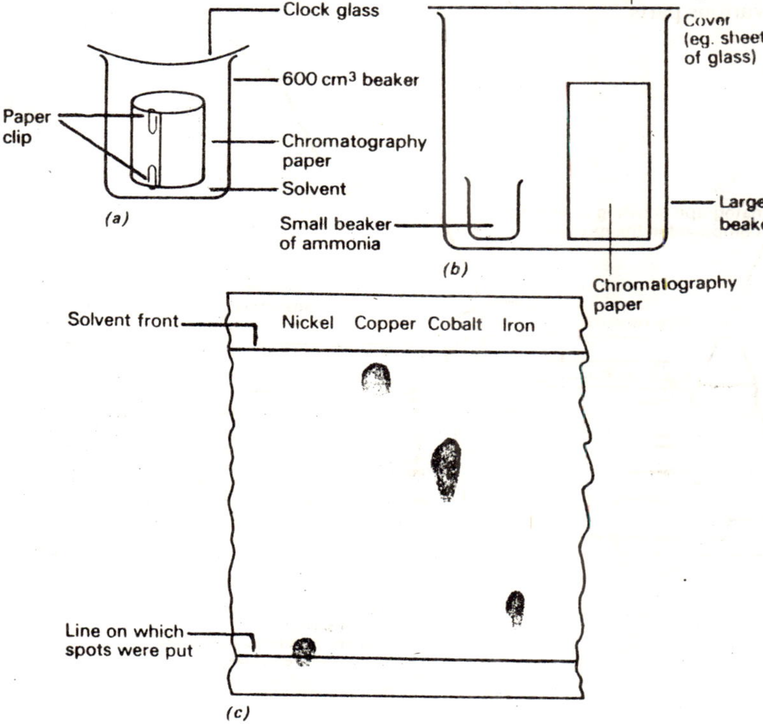

Fig. 1.7. Paper chromatography.

shown in figure 1.7(*c*) should appear. You may find that your samples of iron, cobalt, nickel and copper salts are impure. Look at your chromatograms carefully. Does the nickel (II) sulphate in your laboratory contain any cobalt (II) sulphate, and vice versa ? They often do, because cobalt and nickel occur together in nature and it is very difficult to get them pure. Also, if you used metal paper clips there will be stains round them. What metal are the paper clips made from ?

GAS CHROMATOGRAPHY

The technique of chromatography can also be applied to mixtures of gases, and *gas chromatography* is very widely used in chemical research and industry. Substances that are gases at room temperature can be separated, or more often liquids can be vaporized and then separated. In *gas chromatography* the liquid in which the gases are to be "partitioned" is put onto an inert solid powder support and put into a tube. The gas mixture is blown down the tube by an inert carrier gas (usually nitrogen or helium) and the gases emerging from the end of the tube are analyzed by one of a number of different types of detector. A simple gas chromatography is shown in Fig. 1.8. is a diagram that illustrates the various parts.

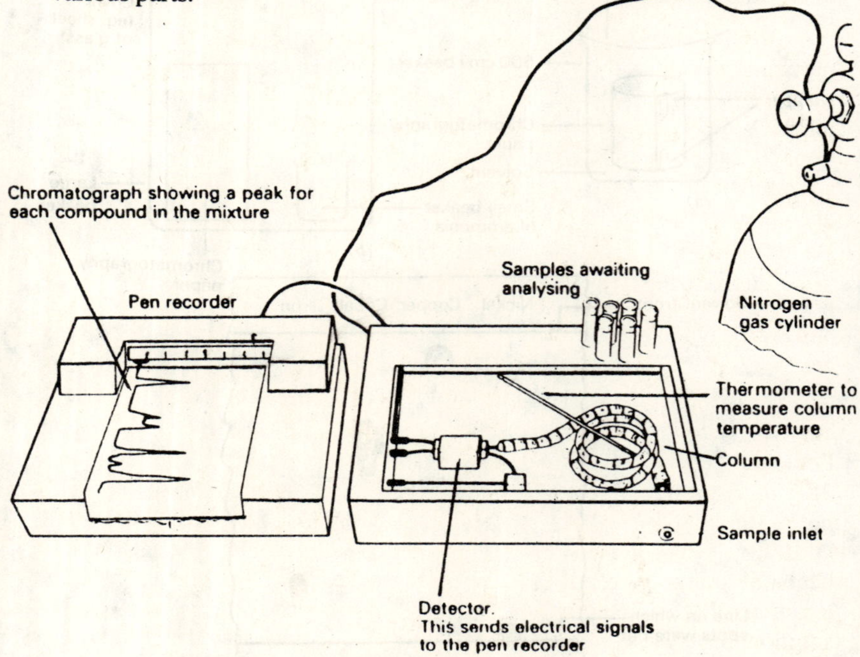

Fig. 1.8. Gas chromatography.

Using gas chromatography, very complex mixtures can be separated and the components identified. Sometimes the sample can be very small (10^{-6} cm^3). Using gas chromatography, hydrocarbons in petroleum can be analyzed, the amount of alcohol in a sample of blood can be measured, the type of pesticide on the leaf of a plant can be determined, and the many constituents is tobacco tar can be identified. These are just four examples of the complex mixtures that can be separated by this technique.

SATURATED SOLUTIONS

If a small amount of salt, or any other soluble solid, is added to water, the salt dissolves. The solution obtained is said to be unsaturated. If further quantities of salt are continually added, then a stage will be reached when no more salt will dissolve; the water is then said to be saturated with the salt. A *saturated solution* is defined as one that contains the maximum amount of solute which, with excess solute present, can be dissolved at the temperature concerned. If the water is heated, more of the solute will probably dissolve because the solubility of most solids increases with temperature.

FRACTIONAL CRYSTALLIZATION

It is possible to measure exactly how much of a solid dissolves in a particular mass of a solvent. Usually the quantity measured is the number of grams of solute required to saturated 100 g of solvent at a particular temperature. These measurements are made by preparing saturated solutions at various temperatures and then analyzing them. The measured solubilities can be plotted against temperature on a graph. Such graphs are called *solubility curves*, and these are given for a few common substances (Fig. 1.9). The graphs show, for example, that in boiling water (at 100° C) the least soluble salt, of those given, is potassium sulphate (25 g/100 g water). Copper (II) sulphate is more soluble (73 g/100 g water), and potassium nitrate is much more soluble than any of the other substances (well over 160 g/100 g water it is off of the top of the graph).

Another useful piece of information that can be obtained from these curves concerns the behaviour of a solution on changing the temperature. For example, if you have a solution in 100 g of water at 100° C containing 25 g of potassium chlorate, then the graph tells us that the solution will not be saturated, as 100g of water will dissolve 53 g of potassium chlorate at 100° C. Now, if the solution is cooled the graph tells us that it will become saturated when the temperature reaches 58° C, and below 58° C less than 25 g of potassium chlorate will be needed to give a saturated solution. So as the solution is cooled further, solid potassium chlorate will be deposited out of the solution until at room temperature, about 18° C, only 7 g will remain in solution; 18 g will have been deposited in the container and could be filtered off.

Now suppose that a addition to 25 g of potassium chlorate, 25 g of sodium chloride was dissolved in the boiling water. The graph tell us that even at room temperature 37 g of sodium chloride are required to give a saturated solution, hence on cooling all of the sodium chloride would remain in solution. By dissolving in boiling water and cooling, therefore, pure potassium chlorate could be obtained from a mixture of potassium chlorate and sodium chloride.

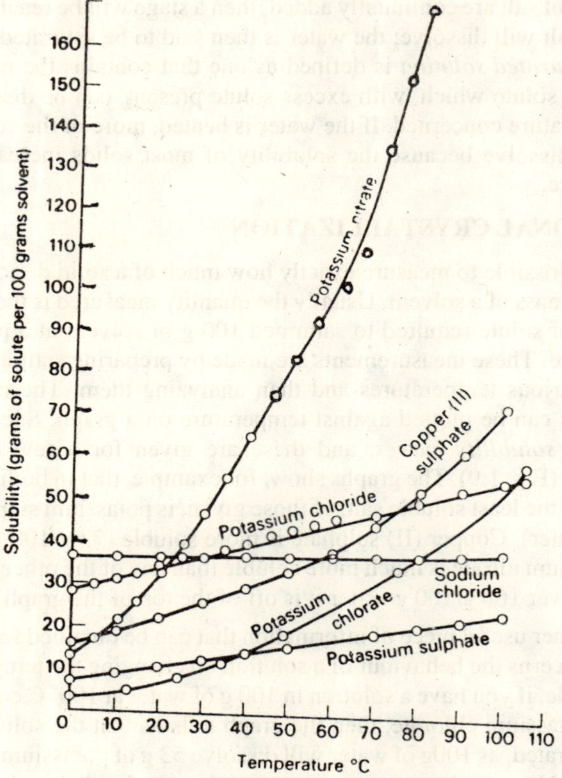

Fig. 1.9. Solubilities of some substances in water

Usually the solubilities of the substances in a mixture are such that a mixture of the substances is deposited on cooling, but the composition of the mixture will be different from that of the solution. This is the basis of a method of separation of a mixture of solids known as *fractional crystallization.* To get substance pure further recrystallization are often necessary. This makes fractional crystallization a very tedious process, but it is a valuable method used for obtaining pure substances from mixtures.

SUPERSATURATION

If hot saturated solutions of some substances are cooled, the expected crystallization does not occur, particularly if the cooling is slow, the solution is left undisturbed, and all dust is excluded. Sodium thiosulphate is one of the few substances that readily shows this phenomenon. If some sodium thiosulphate crystals are gently heated with a very little water they dissolve. The ease of solution is partly due to the high percentage of water of crystallization in the crystal. There is so much, in fact, that it is possible for the sodium thiosulphate to dissolve in its own water of crystallization. On cooling no crystals form.

Such a solution is said to be *supersaturated.* It is in an unstable state. The addition of single crystal of sodium thiosulphate acts as a nucleus on which other crystals can form, and crystallization occurs. There is a noticeable evolution of heat when the crystals form. This is because, when a solvent dissolves a solute to form a solution, heat is usually taken in by the solution. Hence when crystals form the solution is changing back to solvent and solute and heat is evolved.

Supersaturation does not occur if there is excess solute present. A *supersaturated solution* can be defined, therefore, as one containing more solute at a particular temperature than it contains is the presence of crystals of the solution.

SOLUBILITY OF GASES IN WATER

So far we have considered only the solubility of solids in liquids. Solution of gases in water are also of importance. The various gases show great differences in their solubilities. For example I cm^3 of water at 0°C will dissolve approximately 1300 cm^3 of ammonia, but only 0.02 cm^3 of hydrogen. The gases that are readily soluble, such as ammonia, hydrogen chloride and sulphur dioxide, all react with the water, but no reaction occurs with the sparing soluble ones, such as hydrogen, oxygen and nitrogen.

Water will inevitable contain small quantities of gases dissolved in it, as a result of its contact with air. This can be illustrated by filling a large flask and delivery tube with tap water. When the flask is warmed a small amount of gas collects in the test tube. The gas will re-ignite a glowing splint—indicative of a high proportion of oxygen and, therefore, of the fact that oxygen is more soluble in water than is nitrogen.

This experiment also illustrates that, unlike nearly all solids, gases are more soluble in cold water than in hot. In fact, a gas that does not react with water can be removed by boiling.

The presence of dissolved oxygen in water is of immense biological importance, since it is this oxygen that fish extract from water as it passes through their gills, and it is their sole means of respiration.

COLLOIDAL SOLUTIONS

We have seen that salt forms a solution in water that is completely homogeneous, i.e., the individual salt particles cannot be seen and the solution passes through a filter paper. We have also seen the finely divided chalk forms a suspension with water. The chalk particles can be readily seen under a microscope and are removable by filtration.

It is possible to prepare a solution where the particles of the substance dispersed in the liquid are intermediate in size between those of solute particles and the particles of a suspension. Such a solution is called a *colloidal solution.*

If a solution of sulphur dioxide is added to a solution of hydrogen sulphide, until the smell of the latter is removed, sulphur is formed. The sulphur is in colloidal solution and the mixture obtained will pass through a filter paper.

Colloidal solutions prepared in this manner are not very stable and the colloidal particles soon coagulate (i.e., join together) to form larger particles that then settle out from the solution. Some colloidal solutions, carefully prepared and stored, last indefinitely.

Some colloidal solutions have beautiful colours. Colloidal gold is a rich deep red colour. This has been known for many centuries and made use of in the making of red glass for stained glass windows in churches.

When the substance dispersed in water is also a liquid the colloidal system obtained is called an *emulsion*. An emulsion is formed if a little paraffin is shaken with water, but it soon separates. If a little soap solution is also added the emulsion is much more stable. Soap is effective in removing oil and grease because it helps to take them into colloidal solution and so helps them to be washed away.

Milk is an important natural emulsion of fat in water. It is a fairly stable emulsion. But cow's milk in contact with gastric juices forms curds, making it less digestible. For this reason cow's milk is not suitable for young babies. Human mother's milk contains an albumin that prevents this coagulation.

CHAPTER 2

Atoms, Molecules, Mole and Equations

DALTON'S ATOMIC THEORY

Dalton's Theory states that :

(*a*) *Matter is made up of small, individisble particles called atoms.*

(*b*) *Atoms can neither be created nor destroyed.*

(*c*) *The atoms of a particular element are alike in every respect, inlcuding mass. They differ from the atom of any other element.*

(*d*) *When atoms combine they do so in small whole numbers to form "compound atoms", "Compound atoms" formed by such combination are alike in every respect. (These "compound atoms" are now called* MOLECULES.*)*

This theory leads us to a common definition of an atom :

An atom is the smallest particle of an element that can take part in a chemical change.

Both the words "law" and "theory" have now been used. It is essential to understand what is implied by these terms. A *law* is a general statement describing a way in which substances behave; it summarizes a number of isolated facts, which can be directly verfied by experiment. A *theory* provides a generally accepted explanation of some way or other in which substances behave. A theory, once formulated, often leads to the discovery of fresh facts; in time, facts are found which are not in accordnace with a particular theory. It then needs modifying, or may have to be abandoned altogether.

These points are well illustrated by reference to Dalton's theory. It provides. Satisfactory explanations for the laws mentioned earlier in the chapter.

(a) Law of conservation of mass

By Dalton's theory atoms can neither be created nor destroyed. It follows that a chemical reaction can consist only of a rearrangement of atoms, with no possible change in mass, i.e. the Law of Conservation of Mass will be obeyed.

(b) Law of constant composition

By Dalton's theory every molecule of a particular compound contains the same number of each type of atom present. Each atom has a fixed mass. It is therefore obvious that all the molecules of the compound should have the same percentage composition by mass, and that all samples of the compound should have the same percentage composition.

From the sort of result obtained from the experiment described in section 3.3, some idea of *relative atomic masses* may be obtained. Roughly 4 g of copper combine with 1 g of oxygen. If it is assumed (as Dalton did) that copper (II) oxide is composed of atoms with one atom of copper for every atom of oxygen, this means that a copper atom is four times as heavy as an oxygen atom. Atoms do not, however, always combine in the simple ratio of 1 : 1.

(c) Law of multiple proportions

If atoms combine together in *small whole numbers*, as suggested by Dalton, then the Law of Multiple Proportions must be true. For example, suppose two elements *X* and *Y* form two compounds.

Assume that in one compound, one atom of *X* combines with one atom of *Y* and that in the other compound, one atom of *X* combines with two atoms of *Y*.

As Dalton assumed all atoms to be alike in every respect, including mass, then it is obvious that the masses of Y combining with a fixed mass of X are in the ratio 1 : 2.

Whatever *small* number of atoms of X and Y are assumed to combine with one another, a simple ratio will exist for the masses of one of the elements combining with a fixed mass of the other.

ATOMS, MOLECULES AND THEIR SYMBOLS

The *atom* has already been defined as the smallest particle of an elements that can take part in a chemical change, and a *molecule* as the species that is formed when atoms combine.

The substances made up of *atoms* are the elements.

The substances made up of *molecules* are elements if all the atoms are alike, or *compounds* if two or more different atoms are combined together.

RELATIVE ATOMIC MASSES

Atoms are extremely small and the determination of their absolute masses was, in Dalton's time, impossible. We now know for example, that the mass of a hydrogen atom is approximately 1.67×10^{-24} g.

It is therefore usual to refer to the *relative masses of atoms*, and by assuming that atoms usually combine together in the ratio 1 : 1 it was possible, as indicated in the previous section, to obtain these relative masses. Becuase this assumption is by no means always correct, incorrect atomic masses were often assigned to some elements in the early nineteenth century.

Until 1961 the mass of the hydrogen atom was taken as a standard and relative atómic mass defined as the number of times one atom of a particular element was heavier than one atom of hydrogen. In 1961, for reasons that need not concern us here, the definition was changed to : *The relative atomic mass of an elements is the number of times one atom of that element is heavier than one twelfth of an atom of the main isotope of carbon (known as Carbon-12).*

If this second definition is used the relative atomic mass of hydrogen changes from 1.000 to 1.008, so that in fact, whichever definition is used the relative atomic masses are almost the same.

Nowadays the relative atomic masses are obtained using a large expensive instrument known as a *mass-spectrometer.* By shooting the atoms through some complicated electric and magnetic fields, this instrument is able to sort atoms according to this mass, and to measure very accurately the masses of the atoms. For this method it does not matter at all in what ratios atoms combine.

RELATIVE MOLECULAR MASSES

In the water molecule, for example, two atoms of hydrogen combine with one atom of oxygen. The *molecular mass* of the water molecule is the sum of the atomic masses of the atoms, i.e., 2 × (atomic mass of H) + (atomic mass of O) = 2 + 16 = 18. Similarly, hydrogen chloride contains one hydrogen atom to each chlorine atom, hence its molecular mass = (1 + 35.5) = 36.5.

It follows from this and the definition of relative atomic mass that the *relative molecular mass is the number of times a molecule of an element or compound is heavier than one twelfth of one atom of the carbon-12 isotope*, (or one hydrogen atom, to use the old standard).

Molecular mass can also be used for compounds that do not exist as molecules (i.e., exist as ions). In these cases the *molecular mass* is the mass of a "*formula unit*", a formula unit being the simplest statement of the combining ratio of the two ions. Thus the formula unit of sodium chloride is NaCl. This says there are equal numbers of sodium and chlorine units in sodium chloride. The *molecular mass* of sodium chloride is therefore 23 + 35.5 = 58.5

(*Note*. Although *relative molecular mass* is the correct term, it is often abbreviated to *molecular mass*.)

AVOGADRO'S CONSTANT

We have seen that the atomic mass of hydrogen is 1, and that of carbon is 12. This means, as we have just explained, that one atom of carbon is 12 times heavier than one atom of hydrogen. Clearly two atoms of carbon will be 12 times heavier than two atoms of hydrogen, and n atoms of carbon (where n is any number) will be 12 times heavier than n atoms of hydrogen. Therefore, *any mass of hydrogen will have in it the same number of atoms as 12 times that mass of carbon.*

Our standard unit of mass is the kilogram, hence we measure mass in kilograms and grams, i.e., *1 g of hydrogen will have in it the same number of atoms as 12 g of carbon.*

We could, of course, choose any other element and its atomic mass expressed in grams will contain the same number of atoms as 1 g of hydrogen and 12 g of carbon (e.g. 16 g of oxygen, or 23 g of sodium). Furthermore, we can choose any molecule and its molecular mass expressed in grams will contain the same number of molecules as there are atoms in 1 g of hydrogen (e.g. 18 g of water or 36.5 g of hydrogen chloride).

You might ask if it is known what this number is ? The answer is that it is known very accurately. It is, as you might expect, a very big number, 6.02×10^{23}. It is called *Avogradro's constant* after the Italian physicist Amedeo Avogardro, who has contributed a great deal to this aspect of chemisty.

THE MOLE

In the last section we saw that:

(*a*) there are 6.02×10^{23} atoms in the atomic mass expressed in grams of an elements.

(*b*) there are 6.02×10^{23} molecules in the molecular mass expressed in grams of an element or compound.

"6.02×10^{23} atoms or molecules or ions" is such an important quantity to the chemist that it is given a special name *the mole* (abbreviated name : **mol**).

The mole is 6.02×10^{23} particles of anything. Usually the mole is used in referring to atoms or molecules or ions or electrons, e.g. 1 mole of copper atoms contains 6.02×10^{23} atoms, and 1 mole of water molecules contains 6.02×10^{23} molecules. The idea of a mole can be applied to anything although, of course, it is normally applied only to these small particles met with in science.

To give some idea of what an enormous number of particles are contained in a mole, try to imagine a mole of people i.e. 6.02×10^{23} people. If they stood packed tightly together over the whole surface of the earth (land and sea) there would be enough left over to cover another 250 000 000 earths (or if they all stood on one earth on everyone elses' shoulders the columns of people would stretch beyond the moon !). You might like to guess (or if you can, work out) the volume of a mole of footballs. (Answer : They have a volume 5 times that of the earth.)

In contrast we have a mole of hydrogen atoms (6.02×10^{23} of them) in just 1 g, and a mole of water molecules (6.02×10^{23} water molecules) in 18 g or a large tablespoonful (≡ 5 teaspoonsful).

Put another way—there are as many molecules of water in a teaspoonful as there would be footballs in the volume of the earth.

MOLAR MASS

Molar masses of atoms and molecules

We have defined a *mole* (6.02×10^{23} particles), and we have seen that a mole of atoms or molecules is present in the atomic or molecular mass expressed in grams. This mass of any substance that contains a mole of particles is called the *molar mass*.

Thus a mole of sodium atoms (atomic mass = 23) has a mass of 23 g. To put it another way, the *molar mass* of sodium is 23 g mol^{-1} (g mol^{-1} means g per mole).

Similarly a mole of water molecules (molecular mass = 18) has a mass of 18 g. Or we can say that the *molar mass* of water is 18 g mol^{-1}.

(These *molar masses* were previously called *gram-atoms* and *gram-molecules*.)

Molar masses of ionic compounds

Compounds, made up of ions do not contain molecules and the 'particles' referred to in the definition of a mole of such a compound must be clearly specified. This is not usually difficult. We use, the "formula unit". For example, for sodium chloride, which has the formula NaCl (or $Na^+ Cl^-$), the particle is taken to be everything represented in the usual formula. The *molar mass* of sodium chloride is, therefore, (23 + 35.5) g mol^{-1} = 58.5 g mol^{-1}. You will see that 1 mole of sodium chloride (6.02 $\times 10^{23}$ formula units) consists of one mole, i.e. 6.02 $\times 10^{23}$, of sodium ions and one mole, i.e. 6.02 $\times 10^{23}$ of chloride ions.

Similarly the molar mass of sodium sulphate (Na_2SO_4 or $(Na^+)_2SO_4^{2-}$) is (2 × 23) + 32 + (4 × 16) g mol^{-1} = 142 g mol^{-1}. Here 1 mole of sodium sulphate (6·02 $\times 10^{23}$ formula units) consists of 2 moles, i.e. 12.04 $\times 10^{23}$, of sodium ions and one mole, i.e. 6.02 $\times 10^{23}$, of sulphate ions.

To summarise

The *mole* is a *number* (6.02 $\times 10^{23}$) of particles; it does not have any mass. The *mass* of a *mole* of particles is the *molar mass*. This has units of grams per mole (g mol^{-1}).

To emphasize the points that have been developed during the last few pages the following table lists them all for some atoms and molecules.

Atom	*Symbol*	*Relative Atomic mass*	*Molar mass*	*Number of atoms in a mole*
Hydrogen	H	1	1 g mol^{-1}	6.02 $\times 10^{23}$
Oxygen	O	16	16 g mol^{-1}	6.02 $\times 10^{23}$
Carbon	C	12	12 g mol^{-1}	6.02 $\times 10^{23}$
Sodium	Na	23	23 g mol^{-1}	6.02 $\times 10^{23}$
Sulphur	S	32	32 g mol^{-1}	6.02 $\times 10^{23}$
Molecule	*Formula*	*Relative Molecular mass*	*Molar mass*	*Number of molecules in a mole*
Hydrogen	H_2	2	2 g mol^{-1}	6.02 $\times 10^{23}$
Oxygen	O_2	32	32 g mol^{-1}	6.02 $\times 10^{23}$
Water	H_2O	18	18 g mol^{-1}	6.02 $\times 10^{23}$
Hydrogen chloride	HCl	36.5	36.5 g mol^{-1}	6.02 $\times 10^{23}$
Carbon dioxide	CO_2	44	44 g mol^{-1}	6.02 $\times 10^{23}$
Sodium chloride	NaCl	58.5	58.5 g mol^{-1}	6.02 $\times 10^{23}$
Sulphur dioxide	SO_2	64	64 g mol^{-1}	6.02 $\times 10^{23}$

Calculation of molar masses

Molar mass was defined as the mass in grams of a mole of particles. To calculate the molar mass of a compound, it is necessary to know the atomic mass of each of its constituent elements. The *molecular mass* is then the sum of the *atomic masses* of the individual atoms and the *molar mass* is this number of grams of the compound.

Note here that the use of formulae is very helpful. It tells us the number of each of the atoms present in the molecule and it defines the "particle" needed to specify a mole.

Carbon dioxide, CO_2 :

Atomic mass of carbon = 12 (or, C = 12)

Atomic mass of oxygen = 16 (or, C = 16)

Each molecule of carbon dioxide contains one atom of carbon (atomic mass of carbon = 12) and two atoms of oxygen (atomic mass of oxygen = 16). The molecular mass is therefore, 12 + 16 + 16 = 44, i.e. the molar mass of carbon dioxide is 44 g mol^{-1}.

It is convenient to set out such a calculation in this manner :

$$\text{molecular mas} = \underbrace{\overset{\text{C} \quad\quad O_2}{\overline{12 + (2 \times 16)}}} = 12 + 32 = 44$$

i.e., one mole of carbon dioxide has a mass of 44 g and contains 6.02×10^{23} molecules.

Aluminum sulphate. $Al_2\ (SO_4)_3$:

Al = 27; S = 32; O = 16.

With compounds like aluminium sulphate, which have a more complicated formula, the procedure is similar, but it is advisable to remove the brackets first.

$$Al_2\ (SO_4)_3 \quad = \quad Al_2 \quad + \quad S_3 \quad + \quad O_{12}$$

$$\text{molecular mass} = (2 \times 27) + (3 \times 32) + (12 \times 16)$$

$$= 54 \quad + \quad 96 \quad + \quad 192 \quad = 342$$

$$\text{molar mass} = 342 \text{ g mol}^{-1}$$

Copper (II) sulphate crystals, $CuSO_4 . 5H_2O$:

This compound contains water of crystallization (see 11.5), and the formula indicates that five molecules of water are combined with each $CuSO_4$ unit.

$$CuSO_4 . 5H_2O = Cu \quad S \quad O_4 \quad\quad 5H_2O$$

$$\text{molecular mass} = 63.5 + 32 + (4 \times 16) + 5\ (2 + 16)$$

$$= 63.5 + 32 + 64 + 90 = 249.5$$

$$\text{molar mass} = 249.5 \text{ g mol}^{-1}$$

Calculation of percentage composition by mass from formulae

If the preceding paragraphs have been understood, it is easy to see how the percentage composition of a compound may be calculated. Taking water as our example, we find that it has a molar mass of 18 g mol^{-1}. Of these 18 g, 16 g are accounted for by the oxygen content, and the remaining 2 g by the hydrogen.

i.e. $\frac{16}{18}$ of the total mass of water is oxygen,

i.e. $\frac{16}{18} \times 100$ per cent = 88.9 per cent.

And $\frac{2}{18}$ of the total mass of water is hydrogen,

i.e. $\frac{2}{18} \times 10$ per cent = 11.1 per cent.

Examples

(*a*) *Calculate the percentage composition by mass of sodium chloride Na = 23: Cl = 35.5*

NaCl = Na Cl

molar mass = 23 + 35.5 = 58.5 g mol^{-1}

Thus in every 58.5 g of sodium chloride there are 23 g of sodium and 35.5 g of chlorine.

$\therefore$ Fraction of sodium in sodium chloride $= \frac{23}{58.5}$

and fraction of chlorine in sodium chloride $= \frac{35.5}{58.5}$

$\therefore$ *Percentage of sodium* $= \frac{23}{58.5} \times 100 = 39.3\%$

and percentage of chlorine $= \frac{35.5}{58.5} \times 100 = 60.7\%$

(b) *Calculate the percentage composition by mass of lead (II) nitrate. Pb = 207; N = 14; O = 16.*

$Pb\,(NO_3)_2 = Pb\ \ N_2\,O_6$

molar mass = 207 + (2 × 14) + (6 × 16)

= 207 + 28 + 96 = 331 *g* mol^{-1}

Proceeding as in last example :

Percentage of lead $= \frac{207}{331} \times 100 = 62.5\%$

Percentage of nitrogen $= \frac{28}{331} \times 100 = 8.5\%$

Percentage of oxygen $= \frac{96}{331} \times 100 = 29.0\%$

(c) Calculate the percentage of water of crystallization in washing soda crystals, sodium carbonate-10-water, $Na_2\,CO_3 \cdot 10H_2\,O$

$$Na_2CO_3 \cdot 10H_2O = \quad Na_2 \quad C \quad O_3 \quad 10H_2O$$

$$\text{molar mass} = (2 \times 23) + 12 + (3 \times 16) + 10\,(2 + 16)$$

$$= 46 + 12 \;\; 48 + 180$$

$$= 286 \text{ g mol}^{-1}$$

i.e. in every 286 g of washing soda there are 180 g of water.

$$\text{Percentage of water} = \frac{180}{286} \times 100 = 62.9\%$$

DETERMINATION OF FORMULAE BY EXPERIMENT

In the last sections the compositions by mass of compounds have been calculated from the formulae of compounds.

The reverse is obviously possible. If the masses of each element present in a certain mass of a substance is known (or if the percentage composition by mass of a substance is known), *the chemical formula can be calculated.* Chemists have a number of ways of determining experimentally the composition by mass of a compound and it is in this way that *chemical formulae are experimentally derived.*

For example, the percentage composition by mass of a certain oxide of sulphur is; oxygen, 50 per cent; sulphur, 50 per cent. Obviously the percentage of either of these elements depends on

(*a*) the atomic mass of that element,

(*b*) the number of atoms of that element in each molecule.

If the atomic mass of oxygen were equal to that of sulphur it is obvious that there would be one atom of oxygen present for each atom of sulphur. In actual fact, the atomic mass of oxygen is 16 and of sulphur 32, i.e., the atomic mass of oxygen is half the atomic mass of sulphur. For the oxide to contain 50 per cent by mass of each element there must, therefore, be two atoms of oxygen for each atom of sulphur. This means that the simplest formula of the oxide of sulphur is SO_2, the formula were S_2O_4, S_3O_6 (i.e., S_nO_{2n}) it would still give the same percentage composition by mass.

Another oxide of sulphur contains 60 per cent oxygen and 40 per cent sulphur by mass. Knowing the atomic masses of oxygen and sulphur to be 16 and 32 respectively, the formula can again be calculated. By similar reasoning there are $\frac{60}{16}$ oxygen atoms for every $\frac{40}{32}$ sulphur atoms in a molecule of this oxide :

$$\frac{60}{16} : \frac{40}{32} = \frac{60}{1} : \frac{40}{2} = 3 : 1$$

i.e. there are three oxygen atoms for each sulphur atom, giving the simplest formula as SO_3.

In general, we can find the relative numbers of each type of atom in a molecule of compound, and hence its simplest formula, by dividing the percentage by mass, or the actual masses, of the various elements by their atomic masses. Only rarely are the figures so convenient as in the examples quoted above. The following problems should, however, make the method clear.

Examples

(*a*) *A compound contains 29.1 per cent sodium, 40.5 per cent sulphur and 30.4 per cent oxygen. Calculate its formula (Na = 23; S = 32; O = 16.)*

The relative numbers of atoms of each element are obtained thus:

$$\text{Relative number of atom} = \frac{\%}{\text{atomic mass}}$$

$$\text{Percentage of sodium} = 29.1 \qquad \frac{29.1}{23} = 1.27$$

$$\text{Percentage of sulphur} = 40.5 \qquad \frac{40.5}{32} = 1.27$$

$$\text{Percentage of oxygen} = 30.4 \qquad \frac{30.4}{16} = 1.90$$

Therefore the molecules of the compound must contain whole numbers of sodium, sulphur and oxygen atoms in the ratio 1.27 : 1.27 : 1.90. The simplest whole number ratio agreeing with these firgues is probably obvious to you. If not it can be found by dividing all the figures by the lowest one.

$$\text{i.e. } 1.27 : 1.27 : 1.90 = \frac{1.27}{1.27} : \frac{1.27}{1.27} : \frac{1.90}{1.27} = 1 : 1 : 1.5.$$

Therefore the simples whole number ratio is 2 : 2 : 3. (N.B. The ratio is not absolutely exact because the figures given in the question are subject to experimental error.)

From these data we obtain $Na_2\,S_2\,O_3$ as the simplest formula for the compound. Such a formula is known as an **empirical formula**.

An empirical formula may be defined as the simplest formula which can be obtained for a compound using its composition by mass.

(*b*) A compound containing water of crystallization has the following percentage composition: iron; 20.1%; sulphur, 11.5%; oxygen, 23.0%; water; 45.3%. Calculate the empirical formula of this compound. (Fe = 56, S = 32, O = 16, H = 1.)

Here a molecule of water can be treated as a unit, in the same way as an atom. The molecular mass of water is 18.

		Relative number of atoms
Percentage of iron	= 20.1	$\frac{20.1}{56} = 0.36$
Percentage of sulphur	= 11.5	$\frac{11.5}{32} = 0.36$
Percentage of oxygen	= 23.0	$\frac{23.0}{16} = 1.44$
Percentage of water	= 45.3	$\frac{45.3}{18} = 2.52$

This indicates a formula $Fe_{0.36}\ S_{0.36}\ O_{1.44}\ 2.52\ H_2O$. As before, the simplest ratio of atoms becomes obvious by dividing the relative numbers of atoms obtained by the smallest number.

$$0.36 : 0.36 : 1.44 : 2.52 : = \frac{0.36}{0.36} : \frac{0.36}{0.36} : \frac{1.44}{0.36} : \frac{2.52}{0.36}$$

$$= 1 : 1 : 4 : 7$$

Therefore the simplest formula for this compound is $FeSO_4 . 7\,H_2O$

(c) *20.00 g of a compound containing carbon, hydrogen, and oxygen are found to contain 8.00 g of carbon and 1.33 g of hydrogen, and oxygen are found to contain 8.00 g of carbon and 1.33 g of hydrogen. Calculate the empirical formula of this compound. (C = 12, H = 1, O = 16.)*

Since the compound contains only carbon, hydrogen and oxygen, the mass of oxygen in 20.00 g of the compound

= 20.00 - (8.00 + 1.33) g

= 10.67 g

Although we are not dealing with percentages in this compound, the method of solution is the same.

		Relative number of atoms
Mass of carbon	= 8.00 g	$\frac{8.00}{12} = 0.67$
Mass of hydrogen	= 1.33 g	$\frac{1.33}{1} = 1.33$
Mass of oxygen	= 10.67 g	$\frac{10.67}{16} = 0.67$

$$0.67 : 1.33 : 0.67 = \frac{0.67}{0.67} : \frac{1.33}{0.67} : \frac{0.67}{0.67} = 1 : 2 : 1$$

Therefore the empirical formula is CH_2O.

(*d*) *In sections 3.3 and 3.4 it was found by experiment that 1.68 g of a copper oxide contains 1.34 g of copper, that 0.50 g of oxygen combine with 6.45 g of lead to give lead oxide, that 0.61 g of oxygen combine with 6.01 g of lead to give a second lead oxide and that 0.95 g of oxygen combine with 6.09 g of lead to give a third lead oxide. Obtain the formulae of the copper oxide and of the three lead oxides. (Cu = 63.5, Pb = 207, O = 16.)*

Copper oxide

1.34 g of copper combine with (1.68—1.34) g of oxygen

1.34 g of copper combine with 0.34 g of oxygen

Relative number of atoms

Mass of copper = 1.34 g $\quad \frac{1.34}{63.5} = 0.0211$

Mass of oxygen = 0.34 g $\quad \frac{0.34}{16} = 0.0212$

0.0211 : 0.0212 = 1 : 1

i.e. the empirical formula is CuO.

First lead oxide

0.50 g of oxygen combine with 6.45 g of lead

Relative number of atoms

Mass of lead = 6.45 g $\quad \frac{6.45}{207} = 0.0312$

Mass of oxygen = 0.50 g $\quad \frac{0.50}{16} = 0.0312$

0.0312 : 0.0312 = 1 : 1

i.e. the empirical formula is PbO.

We leave you to work out the formulae of the other two lead oxides.

EQUATIONS

An equation is a convenient method for representing a chemical change by using formulae and symbols. We have already represented a number of chemical reactions by using "word equations", and the process of building up equations will be understood by converting a number of these into normal chemical equations.

How to Construct Equations

(a) Mercury (II) oxide → Mercury + Oxygen.

To construct a chemical equation for this reaction, the formula for each substance involved must be known :

Mercury (II) oxide . . . . HgO

Mercury . . . . Hg

Oxygen . . . . O_2

(N.B. Most gaseous elements normally exist as aggregates of two atoms, see section 7.7.)

A provisional equation is then written :

$$HgO \rightarrow Hg + O_2$$

This equation does not "balance", i.e. the numbers of atoms and molecules involved must be so adjusted that the total numbers of each type of atom on either side of the equation are equal. To give two atoms of oxygen, two molecules of mercury (II) oxide are required. These will also give two atoms of mercury. Therefore the equation becomes:

$$2HgO \rightarrow 2Hg + O_2$$

i.e., two molecules of mercury (II) oxide give two atoms of mercury and one molecule of oxygen. Note that the figure 2 in 2HgO applies to the whole molecule, but that when a small figure is used as a subscript, e.g. in $PbCl_2$ and $Pb(NO_3)_2$, it applies only to the atom or ion preceding it. You might be tempted to alter a formula as an easy method of making an equation balance. *This must never be done. A particular compound has a definite and unalterable formula.*

(b) Potassium chlorate → Potassium chloride + Oxygen.

Inserting the formulae for these substances we get :

$$KClO_3 \rightarrow KCl + O_2$$

To balance this equation there must be an even number of oxygen atoms on the left-hand side of the equation (why ?). Therefore it is possible that two molecules of potassium chlorate will be required.

Trying this, we see that two molecules of potassium chloride and three of oxygen will be obtained, and the equation then balances.

$$2KClO_3 \rightarrow 2KCl + 3O_2$$

(c) Iron + Copper (II) sulphate → Copper + Iron (II) sulphate.

Inserting the formulae we get :

$$Fe + CuSO_4 \rightarrow Cu + FeSO_4$$

The equation balance without further adjustment.

(d) Hydrogen + Oxygen → Water.

Inserting the formulae we get :

$$H_2 + O_2 \rightarrow H_2O$$

Two molecules of water will be formed from one molecule of oxygen. This means that two molecules of hydrogen will be required. Therefore the balanced equation is :

$$2H_2 + O_2 \rightarrow 2H_2O$$

(e) Zinc + Hydrochloric acid → Zinc chloride + Hydrogen.

Inserting the formulae we get :

$$Zn + HCl \rightarrow ZnCl_2 + H_2$$

To give one molecule of zinc chloride, two atoms of chlorine, obtainable from two molecules of hydrochloric acid, are required. The equation then becomes :

$$Zn + 2HCl \rightarrow ZnCl_2 + H_2$$

This now balances.

The construction of equations becomes easy with practice, and to help with this your teacher will no doubt provide a number of examples for you to do. Do not try to learn correct equations in a parrot fashion. There are so many equations that have to be known that this method is never successful. A knowledge of the chemical reaction concerned, and of the way equations are built up, is all that is required.

State Symbols

Substances can exist as solids, liquids or gases and all three states can take part in chemical reactions.

Thus in the reaction :

$$2H_2 + O_2 \rightarrow 2H_2O$$

two gases (hydrogen and oxygen) are reacting to form a liquid (water). In the reaction :

$$2HgO \rightarrow 2Hg + O_2$$

a solid (mercury (II) oxide) is reacting to give two gases (mercury vapour and oxygen).

It is often desirable to show the state (solid, liquid or gas) of the substances represented by a chemical equation. This is easily done by using the symbols s, for solid, I, for liquid, and g, for gas, and putting these symbols (known as *state symbols*) in brackets after the formula.

Thus the above two equations becomes :

$$2H_2 (g) + O_2 (g) \rightarrow 2H_2O (I)$$

and

$$2HgO (s) \rightarrow 2Hg (g) + O_2 (g).$$

Very often we study reactions in solution in water (aqueous solutions). For example :

$$Zn + 2HCl \rightarrow ZnCl_2 + H_2$$

is a reaction between solid zinc and an aqueous solution of hydrogen chloride (i.e. hydrochloric acid) to give an aqueous solution of zinc chloride and hydrogen gas. The hydrogen chloride here is not a solid, a liquid or a gas but it is in aqueous solution, and a state symbol for aqueous solution is needed. The symbol used is (aq).

So the equation is :

$$Zn (s) + 2HCl (aq) \rightarrow ZnCl_2 (aq) + H_2 (g)$$

It must be remembered that many substances can exist as solids, liquids or gases depending on their temperature. Hence water may be formed as liquid water, but very often it is formed as steam (gaseous water). Hence the state symbol used may change if the reaction is carried out at a different temperature.

The reaction

$$2H_2\,(g) + O_2\,(g) \rightarrow 2H_2O\,(l)$$

is usually carried out by burning hydrogen in oxygen, and there is no doubt that in the hot flame the water formed is a gas. The above equation is better written :

$$2H_2\,(g) + O_2\,(g) \rightarrow 2H_2O\,(g)$$

Another commonly met situation where the substance formed is not in its usual (i.e. at room temperature) state is when metals are extracted from their ores. Here high temperatures are often used and the metal is produced as a liquid. In such a situation the appropriate equation would be, for example :

$$Fe_2O_3\,(s) + 3CO(g) \rightarrow 2Fe\,(l) + 3CO_2\,(g).$$

Calculations involving equations

(a) Moles

Chemical equations are of immense value to the chemist, as they tell him the combining ratios of different substances and the ratios of the products formed.

Consider the equation :

$$Fe + CuSO_4 \rightarrow Cu + FeSO_4$$

This equation says that *1 atom of iron* reacts with *1 molecule of copper (II) sulphate* to give *1 atom of copper* and *1 molecule of iron (II) sulphate.*

It also says that *1 mole of iron* reacts with *1 mole of copper (II) sulphate* to give *1 mole of copper* and *1 mole of iron (II) sulphate.*

There is 1 mole of each of the reactants and 1 mole of each of the products. *This 1 : 1 ratio does not always occur.*

For example, consider now the equation :

$$2H_2 + O_2 \rightarrow 2H_2O$$

This equation says that 2 molecules of hydrogen react with one molecule of oxygen to give 2 molecules of water, or that *2 moles of hydrogen* react with *one mole of oxygen* to give *2 moles of water.*

The equation always tells us how many moles of each reactant are needed and how many moles of each product are formed. Of course, to be correct the equation must be properly balanced. It is therefore essential for you to learn how to balance chemical equations as we have just described.

(b) Masses

As the equation tells us the number of moles of reactants and of products we can, by using atomic masses, calculate the masses of reactants and products.

Consider the equations :

$2\,HgO$	$\rightarrow 2\,Hg$	$+\ O_2$
2 molecules	2 atoms	1 molecule
2 moles	2 moles	1 mole

The molar mass of mercury (II) oxide is (201 + 16) g mol^{-1} = 217 g mol^{-1}.

The molar mass of mercury is 201 g mol^{-1}.

(2 × 217) g mercury (II) oxide → (2 × 201) g mercury + 32 g oxygen. i.e., 434 g mercury (II) oxide → 402 g mercury + 32 g oxygen.

Examples

(1) *Calculate the masses of each substance formed when 10 g of calcium carbonate react with excess of hydrochloric acid.* (Ca = 40, C = 12, O = 16, Cl = 35.5, H = 1.)

First a balanced equation must be produced and then the molecular masses added :

$$CaCO_3 \quad + \quad 2HCl \quad \rightarrow \quad CaCl_2 \quad + \quad H_2O \quad + \quad CO_2$$

$$40 + 12 + (3 \times 16) + 2\,(1 + +35.5) \rightarrow 40 + (2 \times 35.5) + (2 \times 1) + 16 + 12 + (2 \times 16)$$

$$100 \quad + \quad 73 \quad \rightarrow \quad 111 \quad + \quad 18 \quad + \quad 44$$

i.e., 1 mole (100 g) of calcium carbonate reacts with 2 moles (73 g) of hydrochloric acid to give 1 mole (111 g) of calcium chloride, 1 mole (18 g) of water and 1 mole (44 g) of carbon dioxide.

Therefore, by simple proportion,

10 g calcium carbonate give $\frac{111}{100} \times 10 = 11.1$ g calcium chloride,

$\frac{18}{100} \times 10 = 1.8$ g water,

and $\frac{44}{100} \times 10 = 4.4$ g carbon dioxide.

(2) *What mass of lead (II) nitrate must be heated to give 22.3 g of lead (II) oxide ? (Pb = 207, N = 14, O = 16)*

The equation is :

$2Pb(NO_3)_2$	$\rightarrow$ $2PbO$	+ $4NO_2$	+ O_2
$2[207 + 2[14 + (3 \times 16)]]$	$2(207 + 16)$	$4[14 + (2 \times 16)]$	(2×16)
662	446	184	32

In this problem we are concerned only with the masses of lead (II) oxide and lead (II) nitrate. There was, therefore, no need to work out the molar masses of the other substances.

From the equation

2 moles (446 g) of lead (II) oxide are obtained from 2 moles (662 g) of lead (II) nitrate.

22.3 g of lead (II) oxide are obtained for $\frac{662}{446} \times 22.3$ g of lead (II) nitrate

= 33.1 g of lead (II) nitrate

(3) What is the maximum mass of silver chloride that can be precipitated from 8.5 g of silver nitrate by the addition of hydrochloric acid ? (Ag = 108, N = 14, O = 16, Cl = 35.5).

The equation is :

$AgNo_3$	+ $HCl \rightarrow$	$AgCl$	+ HNO_3
$[108 + 14 + (3 \times 16)]$		$(108 + 35.5)$	
170		143.5	

From the equation :

170 g of silver nitrate yield 143.5 g of silver chloride

8.5 g of silver nitrate yield $\frac{143.5}{170} \times 8.5$ g of silver chloride

= 7.17 g silver chloride

(4) A crystal of Iceland Spar (calcium carbonate), which is insoluble in water, has a mass of 15 g. It is placed in a solution containing 12.6 g of nitric acid. What mass of crystal will remain after reaction has ceased ? (Ca = 40, C = 12, O = 16, N = 14, H = 1).

The equation is :

$CaC)_3$	+ $2\,HNO_3$	$\rightarrow Ca\,(NO_3)_2 + H_2\,O + CO_2$
$[40 + 12 + (3 \times 16)]$	$2\,[1 + 14 + (3 \times 16)]$	
100	126	

From the equation :

126 g of nitric acid react with 100 g of Iceland Spar

12.6 g of nitric acid react with $\frac{12.6}{126} \times 100$ g of Iceland Spar

$= 10$ g of Iceland Spar.

Mass of Iceland Spar remaining = (15 - 10) g = 5 g

CHAPTER 3

The Electronic Structure of Atoms and the Periodic Table

ATOMS

If we take a piece of pure iron and divide it up continuously, a time will come when our tools, however sharp, will not be able to cut the fragments into smaller pieces. However, if we imagine that we have a tool with which we can subdivide the metal much further the question eventually arises 'Is the metal made up of identical particles which cannot be cut up into yet smaller identical particles?'

Atomic theory gives the answer 'yes' to this question, and according to the theory the smallest identical particles are called atoms. For example, it can be calculated that a 10 g sample of pure iron contains $1.1. \times 10^{23}$ iron atoms, each atom being 2.5×10^{-10}m in diameter. The same is true of all materials in the world; for example, the pens we write with, the clothes we wear, the air we breathe and even our bodies, are made up of many millions of atoms held to gather by chemical bonds.

These tiny atoms are not all alike. The atoms of each element have the same mass (weight), whereas atoms of different elements have different masses. (Over one hundred there different elements have been discovered, e.g. hydrogen, sodium, iron, uranium, see Table 3.2.) Our piece of iron contains only one kind of atom, the iron atom. Pure elements are not normally encountered outside chemical laboratories and most substances we use are compounds, made up of a variety of kinds of atom firmly bonded together. For example, water is a compound containing two different types of atom (hydrogen and oxygen) whereas 'Maneb' (a widely used fungicide) contains five different kinds of atom (carbon, hydrogen, nitrogen, sulphur and manganese).

THE STRUCTURE OF ATOMS

The electron

Let us look at how our knowledge about the structure of the atom has developed since the time of John Dalton at the beginning of the last century.

In *1808 John Dalton* developed the idea that a *solid atom* was the fundamental building block of all matter. No further development of the structure of the atom occurred during most of the 19th century.

In *1897* work by *J.J. Thompson* (in which he passed electric discharges through gases) led to the discovery of a small negatively charged particle known as electrons.

In 1909 R. Millikan determined the charge on a single electron and its mass. Approximately 1840 of these electrons have the same mass as one hydrogen atom. Electrons in varying numbers are constituents of all atoms. They are, as we shall see, the most important atomic particle for the chemist, as it is the electrons that interact when chemical reactions take place.

The nucleus, protons and neutrons

Although atoms are the smallest *identical* particle, each atom can be subdivided into a nucleus, which is positively charged, and a number of electrons, which are negatively charged. In turn, the nucleus an be divided up into a number of positive protons and a number of uncharged neutrons. Since atoms are not electrically charged, the number of electrons in an atom is always the same as the number of protons, as shown, for some small atoms, in Table 3.1.

TABLE 3.1. Number of protons, neutrons and electrons in the atoms of a selection of elements.

Element	*Proton*	*Neutrons*	*Electron*
Hydrogen	1	0	1
Helium	2	2	2
Nitrogen	7	7	7
Phosphorus	15	16	15
Potassium	19	20	19

In 1913 Lord Rutherford showed that an atom consists of a positively charged *nucleus* and that most of the mass of the atom was in the nucleus. He gave the name *proton* to the basic particle making up the nucleus having a charge of +1 (the electron being given a charge of –1) and a mass of 1. [The charge and mass values of 1 are arbitary "atomic" units]

Rutherford said that the electrons revolve around the nucleus, like the planets round the sun, and that there must be a large volume of empty space between the nucleus and the electrons. Niels Bohr worked out the orbits of the electrons and was able to calculate energies of electrons that were in agreement with experimental values.

Almost all of the mass of an atom is in the nucleus. For example, in the simplest atom (hydrogen) the nucleus (1 proton) accounts for 99.95% of the total mass, the single electron contributing only 005%. On the other hand, the volume of the nucleus is tiny compared with that of the whole

Table 3.2. Atomic properties of the elements.

Element	*Atomic number*	*Atomic mass*	*Element*	*Atomic number*	*Atomic mass*
Hydrogen (H)	1	1.008	Potassium (K)	19	39.10
Helium (He)	2	4.003	Calcium (Ca)	20	40.08
Lithium (Li)	3	6.94	Scandium (Sc)	21	44.96
Beryllium (Be)	4	9.01	Titanium (Ti)	22	47.90
Boron (B)	5	10.81	Vanadium (V)	23	50.94
Carbon (C)	6	12.01	Chromium (Cr)	24	52.00
Nitrogen (N)	7	14.007	Manganese (Mn)	25	54.94
Oxygen (O)	8	16.00	Iron (Fe)	26	55.85
Fluorine (F)	9	19.00	Cobalt (Co)	27	58.93
Neon (Ne)	10	20.18	Nickel (Ni)	28	58.71
Sodium (Na)	11	23.00	Copper (Cu)	29	63.54
Magnesium (Mg)	12	24.31	Zinc (Zn)	30	65.37
Aluminium (Al)	13	26.98	Gallium (Ga)	31	69.72
Silicon (Si)	14	28.09	Germanium (Ge)	32	72.59
Phosphorus (P)	15	30.97	Arsenic (As)	33	74.92
Sulphur (S)	16	32.06	Selenium (Se)	34	78.96
Chlorine (Cl)	17	35.45	Bromine (Br)	35	79.91
Argon (Ar)	18	39.95	Krypton (Kr)	36	83.80

atom. There are many ways of illustrating this; for example, if the nucleus of a hydrogen atom were enlarged so as to have the diameter of a pinhead (about 1 mm), then the diameter of the whole atom would be about 100 metres. Thus we should imagine the nucleus of an atom as a minute, extremely dense, body inside a large sphere, which is nearly empty apart from a few rapidly circulating electrons.

At this point it should be stressed that an electron from a hydrogen atom is identical to an electron from any other element. The same is true for protons and neutrons. Atoms of one element are distinguished from those of another element by the number of protons in the nucleus (the atomic Number), and not by any difference in the protons (electrons or neutrons) themselves.

The idea that an atom is mainly empty space can be illustrated by comparing the atom with the solar system.

Atoms have a diameter of about 10^{-10} m, and the nucleus has a diameter of 10^{-14} to 10^{-15} m.

If the nucleus were the same size as the sun, the outer electrons would be farther from it than is Pluto—Pluto is forty times as far from the sun as is the earth—there really is a lot of empty space !

In 1932 Sir James Chadwick discovered another particle in the nucleus with a mass similar to that of the proton but with no electric charge. This was called the *neutron.* All nuclei (except hydrogen) contain neutrons as well as protons.

During the 1920's other ideas than those of Rutherford and Bohr were being developed. These ideas, which are not needed by us, have given us better mathematical descriptions of how the electron moves around the nucleus. What is done is to use the mathematical equations of wave-motion to describe how large collections of electrons behave.

The Planetary atom

The atom of any element is composed of two parts:

The nucleus. This is at the centre of the atom and contains *protons (symbol '''p'')* and *neutrons (symbol ''n'').* It is minute in size compared with the atom as a whole, although it supplies nearly all the mass. The number of protons in an atom of a particular element is referred to as its *atomic number,* and the number of (protons + neutrons) as the *mass number.*

A system of *electrons (symbol ''e'')* rotating around the positively charged nucleus.

Since the atom is electrically neutral, the number of electrons must equal the number of protons.

Symbol for atoms

It is very convenient to be able to give the *atomic number* (i.e. number of protons or electrons) and the *mass number* (i.e. number of protons + neutrons) of an atom in a concise way. The method used is very simple. The *atomic number* is written in front of the symbol for the atom and just below it, and the *mass number* is written in front of the symbol for the atom and just above it.

Thus $^{12}_{6}C$ means a carbon atom, that has an atomic number of 6 and a mass number of 12. (6 protons and (12-6) = 6 neutrons.)

$^{35}_{17}Cl$ is a chlorine atom, that has 17 protons and (35-17) = 18 neutrons.

We can now also define an *element* as a *substance consisting entirely of atoms with the same atomic number* (i.e., with the same number of protons or electrons).

Thus the *hydrogen* atom which has an atomic number of 1 and a mass number of 1, contains 1 protons in the nucleus and has 1 electron, and the *helium* atom, with an atomic number of 2 and a mass number of 4, contains 2 protons and 2 neutrons in the nucleus and has 2 electrons.

Isotopes and relative atomic mass

The relative atomic mass is defined as the *mumbler of times one atom of a particular element is heavier than one twelfth of an atom of carbon-12.* ON this basis the relative atomic mass will be the same as the mass number since protons and neutrons have the same mass and the mass of the electrons can be neglected. However, inspection of the table of relative atomic masses shows that this is clearly not so. The reason is that most elements have *isotopes, i.e., they consist of a mixture of atoms that all have the same number of protons but different numbers of neutrons.*

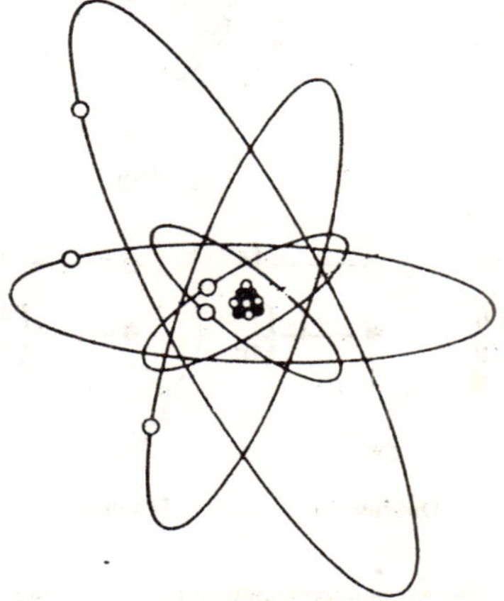

Fig 3.1 The planetary ato....

Hence the atomic numbers are the same (they are, therefore, atoms of the same element) but the mass numbers are different. For example, chlorine exists as two isotopes, chlorine-35 ($^{35}_{17}Cl$) and chlorine-37 ($^{37}_{17}Cl$). The nucleus of each chlorine-35 atom contains 17 protons and 18 neutrons, and that of each chlorine-37 atom contains 17 protons and 20 neutrons (see figure 3.2). There are about 3 times as many chlorine-35 atoms in chlorine gas as there are chlorine-37 atoms. Hence the relative atomic mass of chlorine gas is $\frac{(35 \times 3) + 37}{4} = 35.5$. This is the number in the table of relative atomic masses.

The element carbon is of importance as it is now used as the standard for relative atomic mases. Carbon has two stable isotopes, carbon-12 and carbon 13. The carbon-12 atom has 6 protons and 6 neutrons in it nucleus, and the carbon-13 atom has 6 protons and 7 neutrons in its nucleus. Naturally occurring carbon is mainly carbon-12 (98.89%), the other 1.11% being carbon-13. This means that the relative atomic mass of natural carbon is greater than the standard 12.000 of carbon-12. As you can see from the table of atomic masses that of carbon is 12.01. This is obtained from the percentages of the naturally occurring isotopes:

$$\text{Relative atomic mass of carbon} = \frac{(98.89 \times 12) + (1.11 \times 13)}{100} = 12.01$$

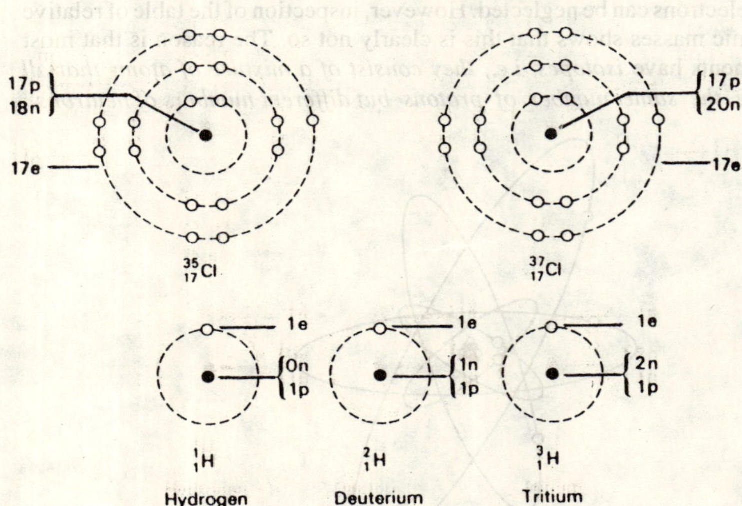

Fig. 3.2 Schematic diagrams of chlorine and hydrogen isotopes showing the electron energy shells.

Three isotopes of the hydrogen atom are known, these are hydrogen-1, hydrogen-2 and hydrogen-3. The hydrogen-1 atom contains 1 proton and 0 neutrons in its nucleus. The hydrogen-2 atom, also known as deuterium, contains 1 proton and 1 neutron in its nucleus, and the hydrogen-3 atom, also known as tritium, contains 1 proton and 2 neutrons in its nucleus (see Fig. 3.2). The natural concentration of deuterium is very low, and that of tritium is almost nil. Hence the relative atomic mass of hydrogen is close to 1 (1.008).

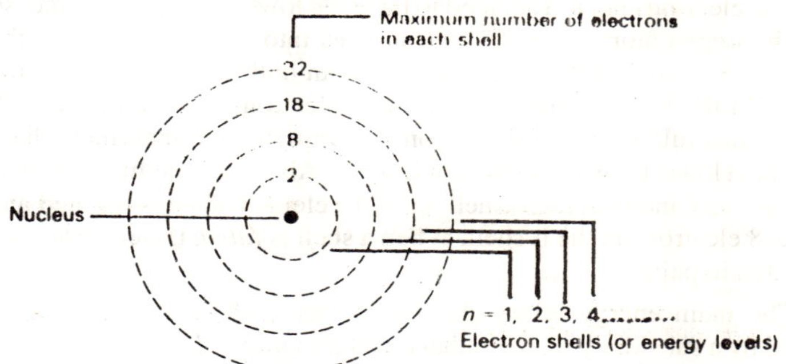

Fig. 3.3 The arrangement of the electron energy shells.

Electron shells and energy levels

(*a*) We have not yet discussed the forces that hold the various parts of the atom together. The most important forces, so far as we are concerned, are those between the charged electrons and protons, i.e. electrostatic forces. It is the electrostatic force of attraction between the positive nucleus and the negative electrons that stops the electrons from flying away from the nucleus.

The electrons of an atom may be regarded as being in concentric rings or shells about the nucleus. These different *electron shells* are different distances from the nucleus. The electrostatic forces *increase* as the distance between the charges *decrease*, hence the electrons that are closet to the nucleus are held most strongly. The more strongly an electron is held the *lower* is its energy, i.e. the more energy you would have to use to pull the electron away from its nucleus. Hence these *electron shells correspond to different level of energy of the electron.*

(*b*) Now each shell can hold up to a certain number of electrons. The first shell, i.e. the one closet to the nucleus, can hold 1 or 2 electrons, the second shell can hold up to 8 electrons, the third shell can hold up to 18 electrons, and the fourth shell up to 32 electrons. (In general the maximum number of electrons in a shell is given by $2n^2$ where n is the number of the shell.) The shells have also been given symbols, these are K, L, M, N.....

Hence:

Number of shell	1	2	3	4
Symbol of shell	K	L	M	N
$2n^2$	(2×1^2)	(2×2^2)	(2×3^2)	(2×4^2)
Number of electrons	2	8	18	32
Energy level	Lowest ——	————	————	Increasing ⟶

The electronic structures of the elements

The electrons in an atom tend to have the lowest energy possible, so in the hydrogen atom the first electron goes into the shell closet to the nucleus (n = 1 or K shell). In the helium atom both electrons go into the K shell. In the lithium atom, the first two electrons go into the K shell which is then full, so the third electron goes into the next innermost shell, i.e. the next lowest in energy ($n = 2$ or L shell). All the succeeding elements fill the K shell and then the L shell, until the element neon is reached and this has 8 electrons in the L shell. When a shell is *filled,* the electrons are associated in pairs.

The main energy levels are divided into sub-levels, each being denoted by a subsidiary or azimuthal quantum number l. Electrons do not really travel in circular orbits. The volume of space where there is a high probability of findings the electron is called an *orbital.* The subsidiary quantum number l describes the shape of the orbital occupied by the electron. l may have values 0, 1, 2 or 3. When $l = 0$, the orbital is spherical and is called an *s* orbital; when $l = 1$, the orbital is dumb-bell-shaped and is called a *p* orbital; when $l = 2$, the orbital is dumb-bell-shaped and is called a *d* orbital and when $l = 3$ a more complicated *f* orbital is formed (Fig. 3.4). The letters *s, p, d* and *f* come from the spectroscopic terms sharp, *p*rincipal, *d*iffuse and *f*undamental, which were used to describe the lines in the atomic spectra.

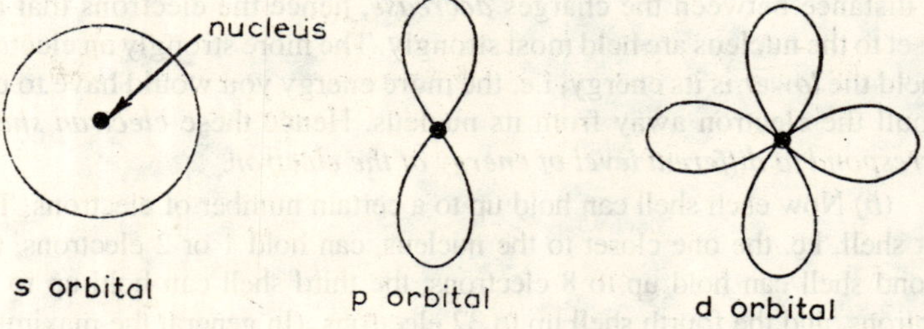

Fig. 3.4 *s, p* and *d* Orbitals

Each main level has a number of sub-levels equal to the principal quantum number. In the first shell of electrons $n = 1$, and there is thus only one value for the subsidiary quantum number. Thus $l = 0$ and this shell contains only *s* electrons.

In the second shell of electrons $n = 2$; hence $l = 0$ and 1, so that both *s* and *p* orbitals exist. Similarly when $n = 3$, $l = 0$, 1 and 2, and *s, p* and *d* orbitals exist.

Each sub-level is further subdivided, the subdivisions being denoted by a magnetic quantum number *m* and a spin quantum number *s*. The magnetic quantum number is determined by the way in which the lines in the atomic spectrum split up under the influence of a magnetic field. *m* can only have the values -1, ..., $-3, -2, -1, 0, +1, +2, +3$, ..., $+l$, where l is the subsidiary quantum number. There are therefore $2l + 1$ values for the magnetic quantum number. The spin quantum number may be regarded as the direction of spin of an electron on its own axis. It has values of $+\frac{1}{2}$ or $-\frac{1}{2}$ corresponding to clockwise and anti-clockwise spin.

In order to show the position of the electrons in an atom, the symbols 1*s*, 2*s*, 2*p*, 3*s*, 3*d*, 4*s*, etc are used to denote the main energy level and sub-level. The electronic structure of some atoms are represented as :

H	$1s^1$
He	$1s^2$
Li	$1s^2, 2s^1$
Be	$1s^2, 2s^2$
B	$1s^2, 2s^2, 2p^1$
C	$1s^2, 2s^2, 2p^2$
N	$1s^2, 2s^2, 2p^3$
O	$1s^2, 2s^2, 2p^4$
F	$1s^2, 2s^2, 2p^5$
Ne	$1s^2, 2s^2, 2p^6$
Na	$1s^2, 2s^2, 2p^6, 3s^1$

THE PERIODIC TABLE

We speak of metallic elements, non-metallic elements, gaseous elements, solid elements, elements that react readily with , say, water, and elements that are inert. All of these classifications follow from the most comprehensive and fundamental method of grouping together the elements–*the periodic table.*

Before periodic classifications, chemistry was a jumble of unrelated facts. During the 18th century the ancient art of *alchemy* had developed to the point where some alchemists (who we can call early scientists) were no longer only concerned with converting other elements into gold but were making careful observations of how different substances behaved. Then in the 19th century, following John Dalton, atomic masses of the elements were being determined.

It was from attempts to relate the chemical and physical properties of the elements and their compounds to their atomic masses that the periodic table was born.

Mendeleev's periodic table

The periodic law

The *periodic law* on which the *periodic table* based was discovered in 1869 almost at the same time an quite independently by a Russian chemist *Dmitri Ivanovich Mendeleev* and a German chemist, *Julius Lothar Meyer.*

It was Mendeleev who appreciated the important general significance of the law that, as stated by Mendeleev, "the properties of the elements are in periodic dependence upon their atomic masses". In his periodic table he left gaps where necessary to ensure that similar elements were grouped together. He also put some elements in positions not justified by the then accepted values for their atomic masses.

The periodic table

Mendeleev's periodic table brings an order to chemistry by collecting together the elements with similar properties. The periodic table illustrated in Fig. 3.5 (see P. 47) is a copy of that published by Mendeleev in 1871. (The original is in German.)

The table is split into horizontal rows that Mendeleev called series (we now call these *periods* as shown on the left of Fig. 3.5. Vertically the table is split into *groups.*

Mendeleev collected together in a group elements with similar properties. By similar properties he meant chemical properties, in particular the ratios in which the elements combine with oxygen or hydrogen. To emphasize this he put the formula of the group oxide and hydride underneath the group number in his periodic table.

THE PERIODIC TABLE

Electronic structure

The different elements have different numbers of electrons and that these are arranged in energy shells. The outermost shells are of highest

Period	*Series*	*Group 1* R_2O_3	*Group 2* *RO*	*Group 3* R_2,O_3	*Group 4* RH_4 RO_2	*Group 5* RH_3 R_2O_5	*Group 6* RH_2 RO_3	*Group 7* *RH* R_2O_7	*Group 8* – RO_4
1	1	H=1							
2	2	Li=7	Be=9.4	B=11	C=12	N=14	O=16	F=9	
3	3	Na=23	Mg=24	Al=27.3	Si=28	P=31	S=32	Cl=35.5	
4	4	K=39	Ca=40	— =44	Ti= –48	V=51	Cr=52	Mn=55	Fe = 56 C0 = 59
	5	(Cu=63)	Zn=65	– =68	– =72	As=75	Se=78	Br=80	Ni = 106 Ag = 108
5	6	Rb=85	Sr=87	?Yt=88	Zr=90	Nb=94	Mo=96	– =100	Ru = 104 Rh = 104
	7	(Ag = 108)	Cd=112	In = 113	Sn = 118	Sb=122	Te=125	I = 127	Pd = 106 Ag = 108
6	8	Cs=133	Ba=137	?Di=138	?Ce=140	–	–	–	– – – –
	9	(–)	–	–		–	–	–	
	10	–	–	?Er=178	?La=180	Ta=182	W=184	–	Os=195 Ir = 197
	11	(Au=199)	Hg=200	Ti=204	Pb=207	Bi=208			Pt = 198 Au = 199
7	12	–	–	–	Th=231	–	U=240	–	– – – –

Fig. 3.5 Mendeleev's Periodic Table of 1871.

Fig. 3.6 The Periodic Table

KEY

Relative atomic mass →	6355	
Atomic number →	29Cu	← Symbol
	Cooper	← Name

1.008 1H Hydrogen

Group → / Period ↓	1	2	←			Transition Metals						→	3	4	5	6	7	0
																		4.003 2He Helium
2	6.939 3Li Lithium	9.012 4Be Beryllium											10.811 5B Boron	12.01 6C Carbon	14.00 7N Nitrogen	16.00 8O Oxygen	19.00 9F Fluorine	20.18 10Ne Neon
3	22.99 11Na Sodium	24.31 12Mg Magnesium											26.98 13Al Aluminium	28.09 14Si Silicon	30.97 15P Phosphorus	32.06 16S Sulphur	35.45 17Cl Chlorine	39.95 18Ar Argon
4	39.10 19K Potassium	40.06 20Ca Calcium	44.96 21Sc Scandium	47.90 22Ti Titanium	50.94 23V Vanadium	52.00 24Cr Chronium	54.94 25Mn Manganese	55.85 36Fe Iron	58.93 27Co Cobalt	58.71 28Ni Nickel	63.55 29Cu Copper	65.37 30Zn Zinc	69.72 31Ga Gallium	72.59 37Ge Germanium	74.92 33As Arsenur	76.96 34Se Ealenium	79.90 35Br Bonnine	83.80 36Kr Kypton
5	85.47 37Rb Rubidium	87.62 38Sr Stronfium	88.91 39Y Yttrium	91.22 40Zr Zirconium	92.91 41Nb Niobium	95.94 42Mo Molybde-num	98.91 43Tc Technetium	101.1 44Ru Ruthenium	102.9 45Rh Rhodium	106.4 46Pd Palladium	107.9 47Ag Silver	112.4 48Cd Cadmium	114.8 49In Indium	118.7 50Sn Tin	121.8 51Sb Antimony	127.6 52Te Tellurium	126.9 53I Iodine	1313 54Xe Xenon
6	132.9 55Cs Caesium	137.3 56Ba Barium	(see below) 57.71	178.5 71Hf Hafnium	180.9 73Ta Tantalum	183.9 75Re Rhenium	186.2 75Re Rhenium	190.2 26Os Osnium	192.2 77Lr Iridium	195.1 18Pt Platinum	197.0 79Au Gold	200.6 80Hg Mercury	204.4 81Ti Thallium	207.2 82Pb Lead	209.0 83Bi Bismuth	12101 84Po Polonium	12101 35At Astatine	12221 86Rn Radon
7	(223) 87Fr Francium	226.0 88Ra Radium	(see below) 89.103															

Lanthanides →	138.9 57La Lanthanum	140.1 58Ce Cerium	140.9 59Pr Praseody-mium	144.2 60Nd Neodymium	(147) 61Pm Promethium	150.4 62Sm Samarium	152.0 63Eu Europium	157.3 64Gd Gadolinium	158.9 65Tb Teribium	162.5 66Dy Dysprosium	164.9 67Ho Holmium	167.3 68Er Erbium	168.9 68Tm Thufium	173.0 70Yb Yiterbium	175.0 71Lu Lutetium
Actinides →	(227) 89Ac Actinum	232.0 90Th Thorium	231.0 91Pa Protactinium	238,0 92U Uranium	237.0 93Np Neptunium	(242) 94Pu Plutonium	(243) 95Am Americium	(247) 96Cm Curium	(247) 97Bk Berkelium	(251) 98Cf Californium	(254) 99Es Einsteinium	(253) 100Fm Fermium	(256) 101Md Mendele-vium	(254) 102No Nobelium	(257) 103Lw Lawrencium

energy, i.e. they are the most reactive. *It is these outermost electrons that take part in chemical reactions and determine the chemical properties of the elements.* It is, therefore, the *electronic structure of the atom* that is the fundamental feature on which the periodic table is based.

A periodic table of the form used these days is illustrated in figure 3.6 (see P. 48). In this we have given the *names* and *symbols* of the elements together with their *atomic numbers* and *relative atomic masses*

The alkali metals

Within a group the elements are similar.

Let us consider the elements of group 1, lithium, sodium, potassium, rubidium and caesium. They are all very reactive metals with a valency of one.

An atom of each of these elements has one electrons in the outer shell, i.e. this electron is of higher energy, and therefore more easily removed, than the others. If it is removed the remaining species is a positively charge ion e.g.:

$$Na \rightarrow Na^{+} e,$$

and the electronic structure now has 8 electrons in its highest energy shell. This, as we have seen, is a particularly stable number of electrons. The size of the atom increases with increase in the atomic number, hence the outer electron moves progressively further away from the nucleus on passing from lithium to caesium. Also the increasing number of inner electrons increasingly screen this outer electron. Because of these two factors it becomes progressively easier to remove the outer electron on moving from lithium to caesium, i.e. *the reactivity of the group 1 elements increases on going down the group.* Therefore potassium reacts more vigorously with water than does sodium, and caesium reacts with explosive violence.

The hydroxides of the metals in group 1 all dissolve in water to give alkaline solutions, hence their name *alkali metals.*

The halogens

Fluorine, chlorine, bromine and iodine are, like the alkali metals, reactive elements with a valency of one. This, too, is expected from their electronic structure.

An atom of each of these elements has seven electrons in its outer shell. They are just one short of the stable octet, and they react by gaining an electron, e.g.:

$Cl + e \rightarrow Cl^{-}$, to give a negatively charged ion.

Like the alkali metals, and the elements in all other groups, the size of the atom increases as the group is descended. The ability to attract electrons is greatest in the small fluorine atom and least with iodine, *hence the reactivity of the group 7 elements decreases on going down the group.*

All of these elements readily form salts with the elements in the early groups (1, 2 and 3). Hence their name *halogens* (salt formers).

The noble gases

Helium, neon, argon, krypton and xenon are five elements that were not known to Mendeleev. They are very unreactive and form very few compounds. This lack of reactivity and the fact that they are gases means that they are found principally in the atmosphere. (Some of these elements are also found with natural gas in certain parts of the world.) Their discovery did not take place until air could be liquefied during the early part of this century, although helium had been observed in the spectrum of the sun some years before.

Their lack of reactivity is readily understood in terms of their electronic structures as they all have a completely full other octet.

Hydrogen

The first element in the periodic table, hydrogen, does not have properties that enable it to be placed satisfactorily in any group. This is due to two unique features. (1) The highest energy shell of a hydrogen atom can hold only two electrons, in contrast to all the others (except helium) that can hold 8 or more. (ii) When hydrogen loses its electron the ion formed, H^+, is a bare nucleus. This is very small in contrast with a positive ion of any other element which must have some electrons still surrounding the nucleus.

Metals and Non-metals

Metals are those elements in the early groups of the periodic table.

Their atoms *tend to lose electrons* to form positive ions.

Non-metals are those elements in the later groups of the periodic table. Their atoms *tend to gain electrons* to form negative ions, rather than to lose electrons.

Metals and non-metals have other characteristics, such as metals being good conductors of electricity and non-metals being poor conductors of electricity.

Metallic characteristics increase as a group is descended. The case of forming a positive ion (a metallic characteristic) increases from lithium to caesium, and the ease of forming a negative ion (a non-metallic characteristic) decreases from fluorine to iodine. If a group near the middle of the periodic table is studied this increase in metallic character on descending group is even more striking.

The transition elements

These elements, do not occur until period 4 and they are all *metals.* They show a close resemblance to one another. Generally they are hard, lustrous and have high melting points. Most of them have several valencies, and their ions in solution are coloured.

CHAPTER 4

The Electronic Theory of Valency and Radioactivity

This chapter consists of two topics. The first, *the Electronic Theory of Valency*, examines the various ways in which molecules can be formed from atoms. The second, *Radioactivity*, looks at the way in which reaction that involve the nucleus of atoms can place.

THE ELECTRONIC THEORY OF VALENCY

How atoms combine

Most of the chemicals that you have met in the laboratory can be divided into two classes, according to their physical properties. Members of the first class of substances are usually crystalline solids. They have high melting points, and, when melted or dissolved in water, allow the passage of an electric current. They do not dissolve in organic solvents. Sodium chloride (common salt), sodium hydroxide (caustic soda), and calcium chloride are common examples.

The other class contains solids, liquids and gases with low melting and boiling points. Only a few of them dissolve in water, and unless there is some sort of combination with the water, as with hydrogen chloride, the solutions do not conduct electricity. None of them will conduct electricity when liquefied. Such substances will often dissolve in organic compounds in general belong to this class. The atoms combine in quite different ways in these two classes, and the compounds formed are said to be *electrovalent* (*or ionic*) and *covalent* respectively.

Ionic or Electrovalent Bonds

Ionic bonds are formed by the *exchange* of electrons between atoms. For example, we can explain how the K and Cl atoms in potassium chloride are held together, in the following way.

The K atom has one electron in its outer shell. In order to achieve the stability of a 'noble gas structure', each K atom needs to lose one electron, thereby forming a charged atom (or ion) K^+, with the electronic structure of Ar. On the other hand, Cl contains seven electrons in its outer shell. To gain a 'noble gas structure', this atom has to gain an additional electron. This will result in a charged ion, Cl^-, which also has the electronic structure of Ar.

It will now be clear that if one electron is transferred from a K atom to a Cl atom, then each atom will achieve a stable electronic state similar to Ar. As a result of this electron exchange, both atoms become charged and the positive K ion (cation) will attract the negative Cl ion (anion). This attraction between opposite charges constitutes an ionic bond.

The simplest way to illustrate ionic bonds is by 'dot diagrams' where the outermost shell electrons are indicated by dots (or crosses) as follows

$$\overset{o}{K} + .\ddot{\underset{..}{Cl}}: \rightarrow \overset{+}{K} + \overset{-}{:\ddot{\underset{..}{Cl}}:}$$

Note that the electrons from K and Cl are indistinguishable; different symbols are used simply to stress the origins of the electrons.

These charged particles, formed when atoms gain or lose electrons are called *ions*. Oppositely charged ions attract one another by *electrostatic attraction*. Electrovalent compounds are, therefore, held together by electrostatic attraction between the oppositely charged ions. The high melting points of these compounds (sodium chloride will just melt in a Bunsen flame) is due to the necessity to break the strong electrostatic attractive forces. The electrostatic forces are more easily broken by water molecules and hence electrovalent compounds often dissolve in water.

Two typical *ionic crystal lattices* are those of sodium chloride (Fig. 4.1) and caesium chloride (Fig. 4.2). In both of these crystals the positive sodium or caesium ions and the negative chloride ions arrange themselves

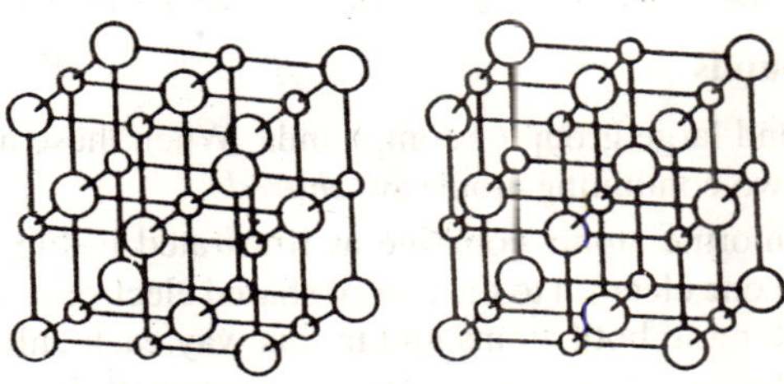

Fig. 4.1 Stereoscopic picture of *sodium chloride*. Either the large or the small circles can be regarded as the sodium (Na^+) ions; the others are the chloride (Cl^-) ions

so that they are equally attracted in all directions. In sodium chloride (Fig. 4.1) each Na^+ ion is in the centre of $6Cl^-$ ions and vice versa. In caesium chloride (Fig. 4.2) each Cs^+ ion is in the centre of 8 Cl^- ions and vice-versa.

It is clear from figures 4.1 and 4.2 that it is impossible to say that a particular sodium (or caesium) ion "belongs" to a particular chloride ion, as they are equidistant between 6 (or 8). This means that "molecules" of an electrovalent compound are not formed. The "molecule" of an electrovalent compound are not formed. The "molecule" is taken to be the atoms in their combining ratio, i.e. given by the formula (NaCl or CsCl).

Electrovalent compounds are formed between *metals* that can attain stable electronic structures by losing one or two electrons and the *halogens* and other species (such as OH^-) that attain a stable electronic structure by having one or two excess electrons. Other ionic compounds, therefore, are magnesium fluoride (Mg^{2+} $2F^-$), Calcium chloride (Ca^{2+} $2Cl^-$), potassium hydroxide (K^+ OH^-), Calcium Oxide (Ca^{2+} O^{2-}) and sodamide (Na^+ NH_2^-).

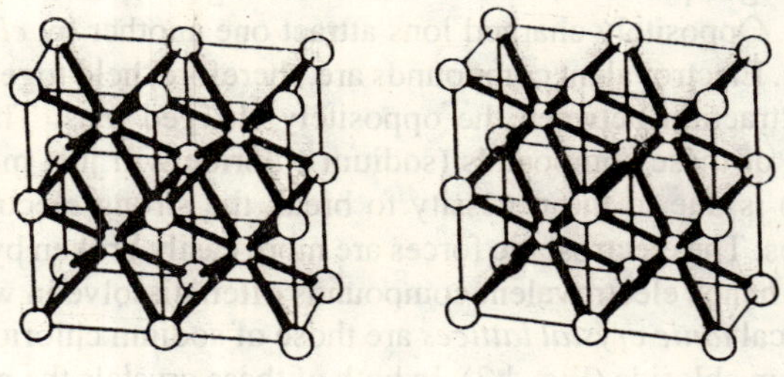

Fig. 4.2 Stereoscopic picture of *caesium chloride*. Either the large or the small circles can be regarded as the caesium (Cs^+) ions; the others are the chloride (Cl^-)ions.

Covalent compounds

This is second large group of compounds. When these are formed electrons on the two combining atoms are *shared*.

Thus two chlorine atoms combine as illustrated in Fig. 4.3, each atom contributing one electron to the pair of shared electrons. The shared electrons then belong to both atoms and in this way each chlorine atom now has eight electrons (the stable octet) in its outer shell. *It is in this way that molecules are formed*. There is little attraction *between* a molecule and its neighbours, and this accounts for the fact that most covalent compounds are gases or liquids.

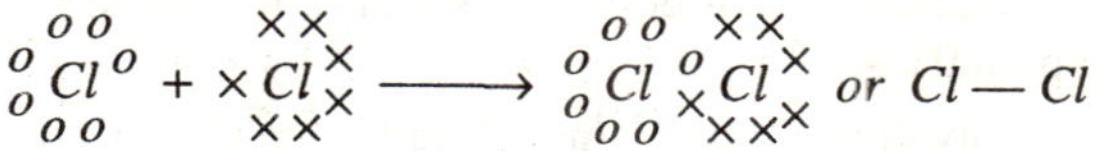

$$Cl + Cl \longrightarrow Cl\,Cl \quad or \quad Cl-Cl$$

Fig. 4.3 Formation of the chlorine molecule.

In certain circumstances, multiple sharing of electrons can occur, resulting in the formation of double or triple covalent bonds, e.g. O_2 and N_2—

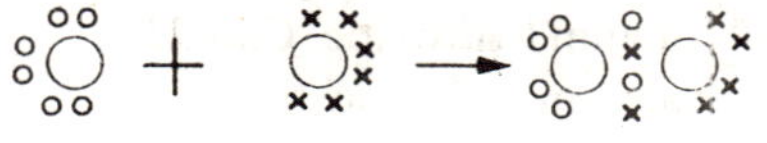

$$O + O \longrightarrow O\,O$$

Double bond
4 electrons
shared

$$or \quad O{=}O$$

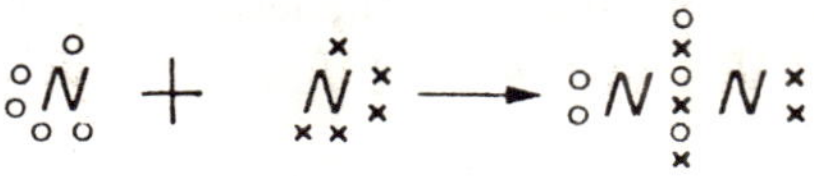

$$N + N \longrightarrow N\,N$$

Triple bond
6 electrons
shared

$$or \quad N{\equiv}N$$

Some Common Ionic and Covalent Compounds

Atoms of the elements on the left of the Periodic table are highly electropositive and therefore tend to form cations, by the loss of one electron (e.g. Na^+, K^+), two electrons (Mg^{2+}, Ca^{2+}) or three electrons (Al^{3+}). On the other hand, atoms of the elements in Group 17 (the Halogens) are highly electronegative and tend to give anions (e.g. F^-, Cl^-, Br^-). Consequently, compounds of metals and halogens (the metal halides) are predominantly ionic (e.g. common salt or sodium chloride (NaCl), potassium chloride (KCl) and a number of less familiar substances such as sodium fluoride (NaF), potassium bromide (KBr) and magnesium chloride ($MgCl_2$).

In these simple ionic compounds, both cations and anions are charged atoms. However, many important ionic substances are made up of polyatomic ions, which consist of a number of atoms cavalently bonded together. For example, in sodium hydroxide, NaOH, the oxygen atom achieves a stable octet of electrons by sharing a pair of electrons with a hydrogen atom (covalent bond) and accepting one electron from a sodium atom (ionic bond)—

$$Na + O + H \longrightarrow Na^{+} + [O\,H]^{-}$$

Thus sodium hydroxide is made up of sodium and hydroxide ions held together by ionic bonds. Since the oxygen and hydrogen atoms are firmly bonded together, the hydroxyl ion does not normally break down in chemical reactions. Consequently, sodium hydroxide is classed as an ionic compound.

Other polyatomic anions, including nitrate (NO_3^-), sulphate (SO_4^{2-}), carbonate (CO_3^{2-}) and different forms of phosphate ($H_2PO_4^-$, HPO_4^{2-} and PO_4^{3-}) occur in ionic substances which are important in agriculture, e.g.

Liming compounds	— calcium hydroxide, $Ca(OH)_2$
	calcium carbonate, $CaCO_3$
Fertilizers	— sodium nitrate, $NaNO_3$
	monocalcium phosphate, $Ca(H_2PO_4)_2$.

There is also one important polyatomic cation, the ammonium ion (NH_4^+), which is the primary product of the mineralization of organic nitrogen in soils.

The coordinate bond

In the last section the *covalent bond* was formed by each of the two atoms that formed the bond *contributing an equal number of electrons to the bond.*

It is also possible to form a covalent bond where *both electrons of the bond are contributed by one atom.* This is called a *coordinate covalent bond.* Consider the two molecules ammonia and boron trifluoride (Fig. 4.4).

$$\begin{array}{ccccccc} \mathrm{H} & & \mathrm{F} & & \mathrm{H} & & \mathrm{F} \\ \mathrm{H:\ddot{\underset{..}{N}}:} & + & \mathrm{\ddot{\underset{..}{B}}:F} & \longrightarrow & \mathrm{H:\ddot{\underset{..}{N}}:} & & \mathrm{\ddot{\underset{..}{B}}:F} \\ \mathrm{H} & & \mathrm{F} & & \mathrm{H} & & \mathrm{F} \end{array}$$

$$\begin{array}{ccccccc} \mathrm{H} & & \mathrm{F} & & \mathrm{H} & & \mathrm{F} \\ | & & | & & | & & | \\ \mathrm{H{-}N:} & + & \mathrm{B{-}F} & \longrightarrow & \mathrm{H{-}N} & \rightarrow & \mathrm{B{-}F} \\ | & & | & & | & & | \\ \mathrm{H} & & \mathrm{F} & & \mathrm{H} & & \mathrm{F} \end{array}$$

Fig. 4.4. Formation of a coordinate bond.

The ammonia molecule has a pair of electrons in the outer shell of the nitrogen atom that are not being used for bonding, and the boron trifluoride molecule has only 6 electrons in the outer shell of the boron. They can, therefore, react as shown in Fig. 4.4 such that the spare pair of electrons on the nitrogen atom is shared with the boron atom. By so doing the nitrogen still has only the octet (3) of electrons in its outer shell, but the boron atom has now also got an octet in its outer shell. The type of bond formed in this manner is a *covalent bond*, but as one of the two atoms has supplied both electrons it is called a *coordinate bond*. The atom that supplies the electrons (here the nitrogen) is called the *donor* atom, and the atom that receives the electrons (here the boron) is called the *acceptor* atom. The coordinate bond is drawn as an arrow (→) *from the donor to the acceptor*.

Other examples of coordinate bonds are in the formation of the ammonium ion from ammonia and a proton (Fig. 4.5).

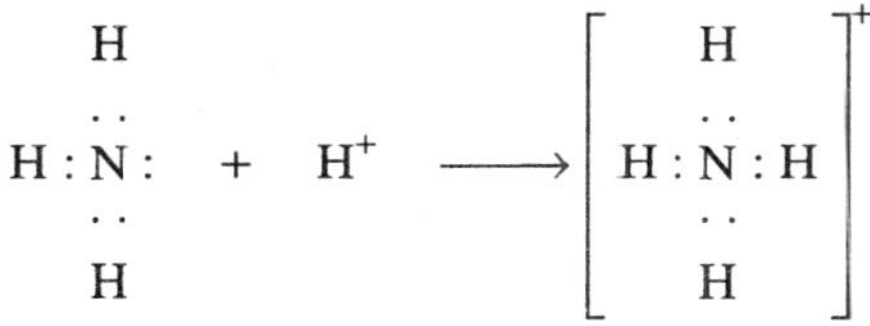

H
|
H—N :   + H^+ ⟶   H—N → H^+
|
H

Fig. 4.5 Formation of the ammonium ion.

The shape of covalent molecules

(*a*) *Introduction*

So far we have not discussed how the atoms in a covalent molecule are arranged in space. Our diagrams have all been 2-dimensional (on a page of the book), but the molecules that we are depicting are 3-dimensional, i.e. all the atoms do not lie in the same plane.

The shape of the simple molecules that we are discussing here can be understood in terms of one very simple idea.

We have already seen that electrons tend to pair together, and we have drawn them so in all our diagrams. The reasons why they form pairs are complicated and it is not possible for us to explain them here. These pairs of electrons are, of course, negatively charged, and like charges repel each other. *Hence each pair repels the other pairs of electrons in a particular shell of an atom as much as possible.*

The most frequently occurring number of electrons in the outer shell of an atom in a molecule is the *octet*. These 8 electrons form 4 pairs and these 4 pairs, which can be regarded as being able to move anywhere on the surface of a sphere surrounding the nucleus, repel each other. They must remain in the spherical shell, and the 4 pairs move to the position of the lowest energy. This position of lowest energy is when they are as far away from each other as possible, i.e. when *the pairs of electrons are at the 4 corners of a regular tetrahedron* (Fig. 4.6).

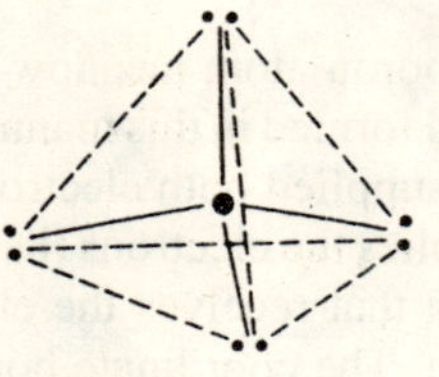

Fig. 4.6 Four pairs of electrons arranged tetrahedrally.

(*b*) *Methane*

Consider, therefore, the *methane* molecule (Fig. 4.7 (*a*)). The carbon atom has 4 pairs of electrons around it; these will be at the corners of a regular tetrahedron. The 4 hydrogen atoms are attached to these 4 pairs of electron, hence the 4 hydrogens are at the corners of a regular tetrahedron with the carbon atom at the centre. The shape of the methane molecule is that of a *regular tetrahedron*.

(*c*) *Ammonia*

Now consider the *ammonia* molecule (figure 4.7(*b*)). The nitrogen atom has 4 pairs of electrons around it and, like those on carbon in methane, these 4 pairs of electrons will be at the corners of a regular tetrahedron. Only 3 of these electron pairs have hydrogen atoms attached to them, as in figure 4.7(*b*). The fourth pair of electrons is known as an unshared (or lone) pair. *Now the shape of the molecule is the shape given by the positions of the nuclei of the atoms making up the molecule*. Hence the shape of the ammonia molecule is that of the three hydrogen and one nitrogen nuclei, i.e. it is a triangular pyramidal molecule.

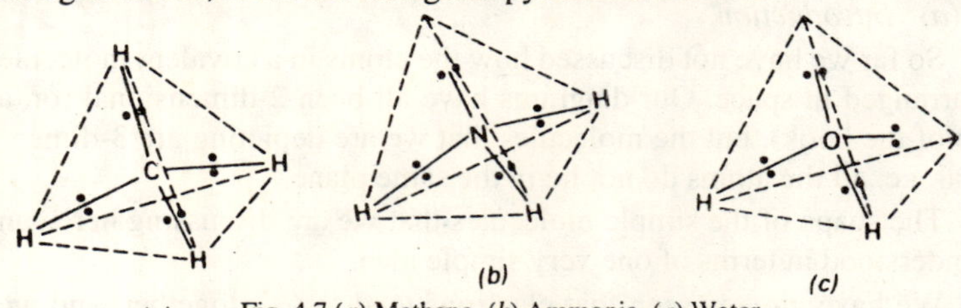

Fig. 4.7 (*a*) Methane, (*b*) Ammonia, (*c*) Water.

(*d*) *Water*

Now consider the *water* molecule (Fig. 4.7(*c*)). The oxygen atom has 4 pairs of electrons around it and, like those of carbon and nitrogen, these 4 pairs of electrons will be at the corners of a regular tetrahedron. Now only 2 of these electron pairs have hydrogen atoms attached to them, as in figure 5(*c*). The water molecule has two lone pairs. The shape of the water molecule is that of the two hydrogen and oxygen nuclei, i.e. it is an *angular* (*V-shaped*) molecule.

(*e*) *Boron trifluoride*

Here, in contrast to methane, ammonia and water, there are only 6 electrons (3 pairs) in the outer shell of the boron atom. These three pairs move to the position of lowest energy (i.e., as far away from each other as possible but keeping the same distance from the nucleus). This is to the corner of a plane triangle (figure 4.8(*a*)). Each pair of electrons is shared with a fluorine atom (Fig. 4.8(*b*)). Hence boron trifluoride is a *plane triangular molecule.*

(*f*) *Molecules containing double and triple bonds*

Methane, ammonia, water and boron trifluoride contain only single bonds. If a molecule contains *double bonds* then the two pairs of electrons making up the double bond together the other pairs of electrons as much as possible. Hence if an atom forms 2 single and 1 double bond to 3 other atoms (as, for example, carbon in *methanol* ($\genfrac{}{}{0pt}{}{H}{H}\!\!>\!C=O$)) the 3 sets of electrons (2 pairs and 1 quartet) repel one another as much as possible. This is exactly what happened in boron trifluoride (figure 4.8); the electrons move to the corners of a plane triangle. *Methanol* is, therefore, a *plane triangular* molecule (Fig. 4.9(*a*)).

Consider, ethane ($CH_2 = CH_2$). Here each carbon atom, like the carbon in methanol, has 2 single bonds, to the hydrogen atoms, and a double bond. Each carbon atom in ethane is, therefore at the centre of a plane triangle, i.e. the ethane molecule is planar (Fig. 4.9(*b*)).

Consider, now, carbon dioxide ($O = C = O$). Here the central carbon uses 4 electrons to bond each oxygen atom. The 8 electrons are, therefore, in two quartets. These two quartets of electrons repel each other as much as possible. You should easily see that they will move to opposite sides of the carbon atom, so *carbon dioxide* is a *linear* molecule (Fig. 4.9(*c*)).

If a triple bond is formed then a sextet of electrons together repel the other pairs of electrons (in an octet, of course, there can only be one other pair). This is the case in *ethyne* ($H — C \equiv C — H$) where the sextet of

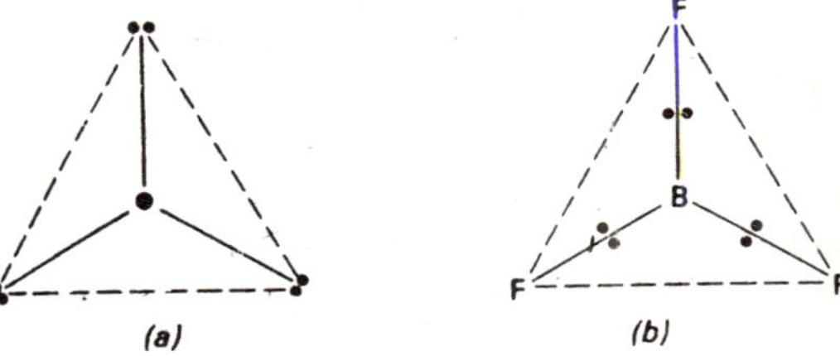

Fig. 4.8 (*a*) Three pairs of electrons arranged in a plane triangle, (*b*) Boron trifluoride.

electrons forming the triple bond repel the pairs of electrons forming the C—H bonds and hence the C—H and C ≡ C bonds will be on opposite sides of each carbon atom. *Ethyne* is, therefore, a *linear* molecule (Fig. 4.9(*d*)).

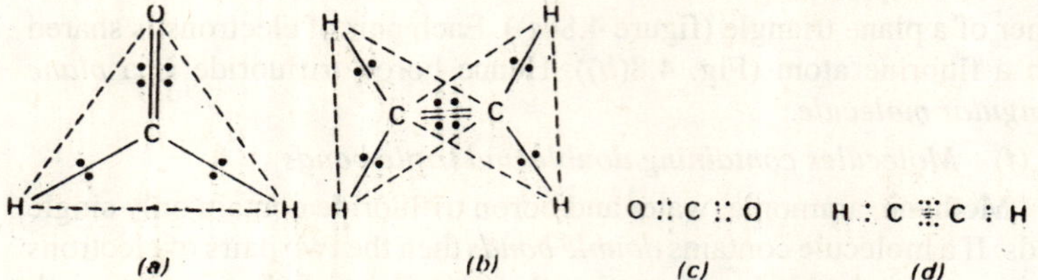

Fig. 4.9 (*a*) Methanol, (*b*) Ethane, (*c*) Carbon dioxide, (*d*) Ethyne.

Polarized Covalent Bonds and Hydrogen Bonds

Consider the covalent using chlorine as an example

In this case the electronegativity of the two participating atoms is the same and the distribution of electrons between the two atoms will be uniform. However, in a covalent bond between atoms of different electronegativities, the shared electrons tend to be displaced towards the more electronegative atom, giving a polarized covalent bond. For example in the OH bonds of the water molecule (H_2O), where the electronegativities of H and O are 2.1 and 3.5 respectively the bonding electrons are slightly displaced towards the O atom, giving a partial separation of charge (where δ indicates a fractional charge). Thus we should think of the covalent bond between an oxygen and a hydrogen atom as having some ionic character.

Overall, purely covalent bonds are formed only between atoms of tne same, or very similar, electronegativity (e.g. C—C, C—H bonds), whereas purely ionic bonds are formed only between highly electropositive metals and highly electronegative non-metals.

Because of the polarization of the O—H bond in water molecules, the partial negative charge on the O atom of each molecule will be attracted to the partial positive charge on the nearest neighbouring water molecule, giving a network of *hydrogen bonds* between the molecules of liquid water—

In more general terms, a hydrogen bond forms between a hydrogen atom (carrying a fractional positive charge) on one molecule and a more electronegative atom (not necessarily oxygen, carrying a fractional negative charge) on another molecule.

As we shall see in Chapter 9, the hydrogen bonds of liquid water are of crucial importance for the existence of life on Earth. They are also important in maintaining the shapes and activities of vital biochemical molecules such as enzymes and nucleic acids.

Molecular Solids

Iodine

Molecules that contain covalent bonds are small molecules as we have just described. These will often be gases, but they do form liquids and solids.

When a solid is formed from small covalent molecules, the molecules are held together in the solid by weak bonds due to the (electrostatic) attraction between the electrons and nuclei of different molecules. A good example is iodine. Solid iodine is composed of I_2 molecules. These pack together in *layers* as shown in Fig. 4.10. This

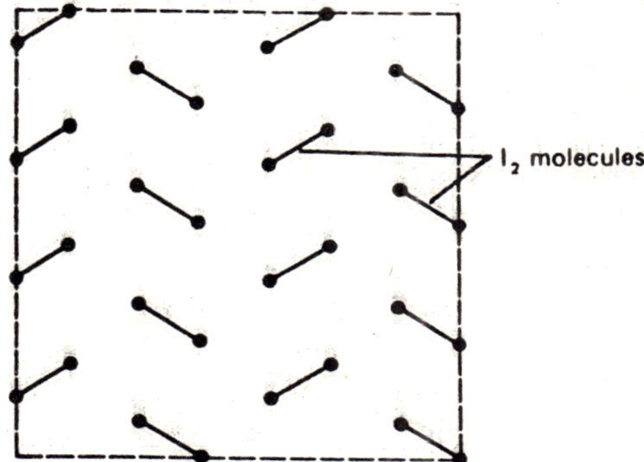

Fig. 4.10 A layer of iodine molecules.

shows the herring-bone pattern of a layer. The iodine molecules are further (by about $1\frac{1}{4}$ times) from molecules in the layers above and below them than they are from molecules in their own layer. These features give solid iodine the characteristic layer-like appearance.

Ice

The importance and widespread occurrence of water means that *ice* is a particularly common solid. When water freezes the water molecules arrange themselves in the ice crystal such that each oxygen atom is surrounded by 4 hydrogen atoms. In this way 6-membered rings of oxygen atoms, with hydrogen atoms in between them, are formed (Fig. 4.11). It is because ice is made up of these 6-membered rings that the snow crystals, illustrated in figure 1.1, are hexagonal in shape.

The electrons of the hydrogen-oxygen bond of the water molecule are, attracted slightly more by the oxygen than the hydrogen. This means that water is a *polar molecule*, i.e. the molecule is like a little bar-magnet (except that is electrically charged and not magnetically charged). The electrostatic attractions that can result from this polarity are the forces that orientate (or arrange) the water molecules as they are found in the ice crystal. (This electrostatic attraction involving hydrogen is in more advanced chemistry books called *the hydrogen bond.*)

One very important consequence of the orientation that takes place when water freezes to ice, is that the solid (ice) has a larger volume and

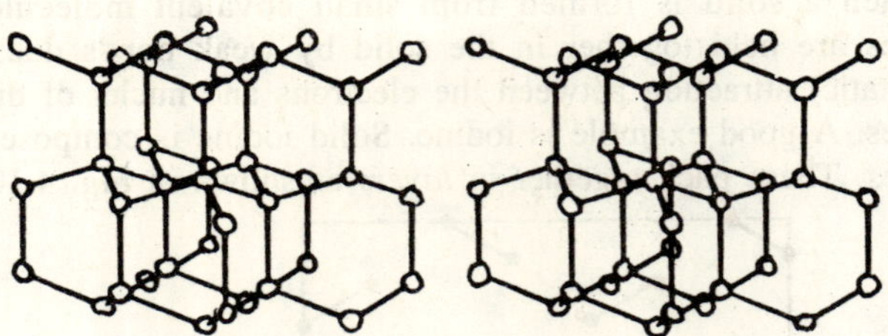

Fig. 4.11 Stereoscopic picture of *ice*. The atoms shown are the oxygen atoms of the water molecules. There is a hydrogen atom in between each pair of oxygen atoms.

hence smaller density than the liquid (water), i.e. *water expands on freezing*. This is most unusual; almost all other liquids contract on solidifying. The fact that water expands on freezing can be troublesome, as you will know if you have ever had a water pipe at home burst during the winter. It also means that ice floats on ponds and rivers, i.e. they freeze from the top downwards.

Macromolecules (giant covalent molecules)

Introduction

Compounds containing covalent bonds are gases, liquids or solids with low melting and boiling points. This is usually the case, but there is also another group of molecules containing covalent bonds where the melting and boiling points can be very high. Indeed the compounds with the highest known melting and boiling points come into this class. These are the *macromolecular covalent solids*.

In a macromolecular solid the covalent bonding extends throughout space so that one can consider *the whole crystal to be one big molecule.*

Diamond and graphite

Diamond is the best example of a macromolecular solid. In diamond (Fig.4.12) each carbon atom is linked by strong covalent bonds to four other carbon atoms arranged tetrahedrally around it. This arrangement extends throughout the whole diamond lattice, and the structure is therefore very rigid, and the bonds are difficult to break. Hence diamond has a very high melting point.

The other allotrope of carbon, *graphite* (Fig. 4.13) is, in contrast, a much softer material. This is because the strong covalent bonding is no longer in three-dimensions but only in two. There are hexagonal layers of carbon atoms held together by covalent bonds, but between the layers are much weaker bonds, and the carbon-carbon distance between layers is much bigger than in a layer. It is for this reason that graphite has a layer-like appearance, and can be split into thin sheets.

Metallic bonding

Another large class of substances is the metals. They contain a further type of chemical bond, the *metallic bond.*

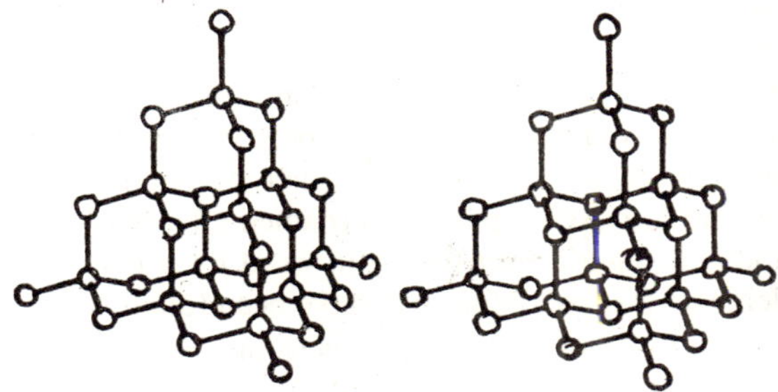

Fig. 4.12 Stereoscopic picture of *diamond.*

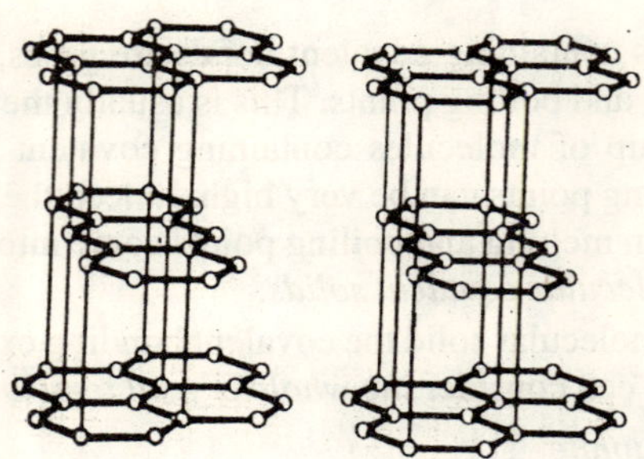

Fig. 4.13 Stereoscopic picture of graphite.

RADIOACTIVITY AND NUCLEAR REACTIONS

John Dalton thought that atoms were indivisible. We now know that this is not correct; it is possible to split up the nucleus of atoms. It should be emphasized however that during chemical reactions it is only the electrons that are involved in reactions. The reactions of nuclei, which is what this section is about, is often called *nuclear physics*.

Usually to break up nuclei you need to put in a lot of energy, but this is not always so; some atoms just break up of their own accord. This phenomenon is known as *natural radioactivity*.

Natural radioactivity

Atoms of certain elements, mainly some of the heavier elements such as uranium, radium and polonium, break up spontaneously and in so doing emit rays and particles. The atom *disintegrates*. This discovery was made in 1896 by the Frenchman Henri Becquerel. He found that photographic plates wrapped in black paper were darkened by uranium salts, and he suggested that the uranium salts emitted some form of rays that affected the film.

In 1898 Marie Curie found that uranium minerals such as pitchblende had more radioactivity than expected from the amount of uranium present. She therefore looked for other radioactive elements in the pitchblende and discovered polonium (named after her homeland of Poland) and radium (named after the Latin word for a ray). Since then many other elements that are radioactive have been found.

The radioactivity in all cases originates from reactions occurring in the nuclei of atoms of the elements concerned. This is different from all *chemical reactions* as discussed elsewhere in this book. These involve only the electrons which surround the nucleus.

The rays resulting from radioactivity are of three types:

(*a*) α (*alpha*) *particles*. These are positively charged helium atoms, He^{2+}, i.e. helium atoms without their two electrons. As they are charged they will be deflected in a magnetic field.

(*b*) β (*beta*) *particles*. These are fast moving electrons. As these have a much smaller mass than α-particles and as they are negatively charged, they are deflected more in a magnetic field and in the opposite direction to α particles.

(*c*) γ (*gamma*) *rays*. These are a form of radiant energy. They are unaffected by a magnetic field (see Fig. 4.14).

When atoms of elements disintegrate to form α, β and γ rays, atoms of new elements are left behind. These may also be radioactive and themselves disintegrate to atoms of other elements. The rate at which atoms disintegrate depends on how much radioactive material is present. As less and less is left, the rate of disintegration becomes slower and slower. A sample of radium, for example, will disintegrate to half its size in 1600 years. What is left will, it its turn, disintegrate to half its size in a further 1600 years, and so on. Thus only after many such periods of 1600

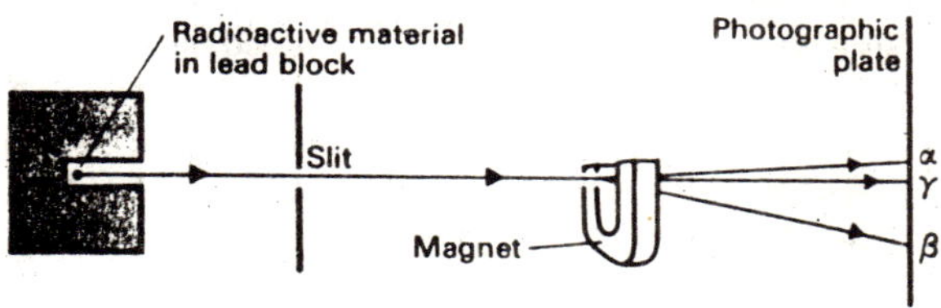

Fig. 4.14 Effect of a magnetic field on α, β and γ rays. (The apparatus is assembled in a vacuum to prevent deflection of rays by gas molecules. The rays that are not emitted forwards are absorbed by the lead block. Three dark spots are produced on the plate showing that γ-rays are not deflected and that α- and β- rays are deflected to different extents in different directions).

years will the sample of radium have completely disappeared. The graph (Fig. 4.15) should make this point clearer.

As the total life of a radioactive element is therefore infinite, it is usual to describe its rate of decay by means of its half-life, i.e. the time taken for half of the element to decompose. Half-lives vary considerably, e.g. radium-226 has a half-life of 1600 years, radon-222 of 3.8 days, uranium-238 of 4.5×10^9 years and polonium-212 of 3×10^{-7} seconds.

Applications of radioactive elements

Radioactive elements are produced, as we have seen, by fission. Some occur naturally, such as radium, and many others can be produced by neutron bombardment of materials put inside a nuclear reactor.

Such radioactive elements have found such numerous uses in industry, medicine, and scientific work, that it is not possible to give them all. Some examples are given below.

Industrial applications

(*i*) *Thickness and density measurements*

Beta-particles can penetrate thin sheets of metal. The proportion of particles absorbed will for a given material be proportional to its thickness; or for a given thickness be proportional to the density of the material. Beta-particle thickness gauges are used for checking the thickness of metal sheets in steelmills and for checking that packets, tins and other containers are correctly filled.

(*ii*) *Tracing pipes and finding leaks*

If the path of water pipe, say, needs to be determined, or a leak detected a radioisotope can be put into the pipe and its passage followed by means of a detector above the ground.

Use in medicine

The best known use of radioactive elements is in the treatment of cancer. Some cancerous growths can be eliminated by exposure to gamma rays. Originally radium was used for this purpose, but now cobalt-60 and caesium-137 are used. Other radioactive isotopes can be implanted in cancerous growths to destroy them.

Scientific application

(*i*) *Study of biological synthesis*

To help understand photosynthesis and related processes, carbon dioxide containing radioactive carbon $^{14}_{6}C$ is used, and the path of this carbon can be followed during the growth of the plant. The chemicals that incorporate the carbon-14 can be isolated and identified.

To study the absorption of phosphate fertilizer by plants, radioactive phosphorus can be used in a similar manner.

Radioactive isotopes, used as *tracers*, are probably the most revolutionary research tools in ecology and agriculture.

For example, we can expose an actively photosynthesizing leaf to 'labelled' carbon dioxide, in which a small proportion of the carbon atoms are $^{14}_{6}C$ instead of $^{12}_{6}C$. As absorption and assimilation of carbon dioxide proceed, tracer carbon atoms (which to the plant are indistinguishable from $^{12}_{6}C$ atoms) become incorporated into organic molecules and it is possible to reconstruct the series of biochemical reactions of photosynthesis by counting the $^{14}_{6}C$ content of a variety of compounds at different times. Over longer periods of time, the translocation of photosynthate from the site of synthesis in the leaf to the site of use or storage in the plant can be charted by monitoring the radiocarbon content of different plant tissues.

Tracers are employed in a wide range of environmental investigations; for example in studies of:

(i) the movement of marine and freshwater sediments;

(ii) the fate of fertilizers, pesticides and pollutants in soils, plants and water;

(iii) the passage of substances (nutrients, drugs, etc.) through large animals (avoiding the wasteful slaughter of animals involved in destructive sampling);

(iv) the age of organic materials (e.g. peat deposits); this application is based on the fact that the natural $^{14}_{6}C$ content of the peat material at the time of deposition falls in a predictable way, according to the half-life, i.e. 50% lost after 5730 years, 75% after 11 460 years, 87.5% after 17 190 years, etc.

With the development of nuclear power stations, radioactive isotopes are also becoming important as environmental pollutants.

In experimental work, synthetic radioactive isotopes, prepared in radiochemical laboratories, are normally used. These laboratories offer a wide range of short- and long-lived isotopes, as well as many organic compounds labelled with $^{14}_{6}C$ and tritium for biological and biochemical studies.

Dating of archaeological specimens

This is another fascinating use of radiochemical techniques. Carbon dioxide in the atmosphere contains a tiny proportion of radioactive carbon-14. In the atmosphere the proportion of carbon-14 remains approximately constant because although it is decaying, it is also being produced by cosmic-rays coming into the atmosphere from outer space. When they are growing the plants, e.g. trees, use in photosynthesis the radioactive carbon in the proportion that it occurs in the carbon dioxide. When growth stops no more carbon dioxide is used, and the radioactive carbon, that is now incorporated in the plants, decays. Carbon-14 has a half-life of 5600 years, and therefore if a sample of preserved wood has half as much carbon-14 in it as a living sample of wood, then its age is

5600 years. If it has a quarter as much its age is 11200 years. If the carbon-14 content can be determined it is quite simple mathematically to determine the age of the sample. Many datings are now made this way, and in general they agree with those made by other methods.

CHAPTER 5

States of Matter

Chemical substances can exist in three different states—solid, liquid and gas. In the chapters 6,7 and 8 we shall look at some important features of each state of matter but before we can do that we must establish the differences between the three states of matter.

If we place a stone (solid state) in a glass beaker, it occupies some of the volume of the beaker but does not change in shape. On the other hand, if we pour some liquid water into the beaker, the water changes in shape to fit the volume. Finally, if we start with a vacuum in the beaker and then admit some hydrogen gas, the gas shapes itself to the beaker as does water, but it does not have a definite volume limited by a surface. Instead, the hydrogen gas expands to fill the beaker and the molecules at the top of the beaker mix freely with the air of the room.

These observations allow us to define each state :

(*i*) *Solids* have definite volumes and shapes, bounded by fixed surfaces ;

(*ii*) *Liquids* have definite volumes but indefinite shapes, bounded by flexible surfaces ;

(*iii*) *Gases* have indefinite volumes and indefinite shapes. Volumes of gases do not have surfaces.

Changes of State

Most substances can exist in each of the solid, liquid and gaseous states. For example, water may be ice, liquid water or water vapour, depending upon the conditions. Normally we are familiar with a substance in one state only (solid iron, liquid mercury, gaseous oxygen) but each substance can exist in other states. If we place solid iron in a sufficiently hot furnace, it will become liquid (melt) ; if we warm mercury in an open vessel, it will become gaseous (vapourize) ; if we expose oxygen to extreme cooling, we obtain liquid oxygen by condensation. These events (melting, condensation, etc.), which are called changes of state, occur at constant temperatures if the substance is pure.

The temperature of a body (solid, liquid or gas) is a measure of the velocity of movement of the atoms or molecules of the body, i.e. the kinetic energy of the particles. We can increase the kinetic energy of these particles by supplying energy in the form of heat ; the more heat we supply, the faster the particles move and the higher will be the temperature of the body.

In the solid state, the atoms of a piece of iron, for example, are tightly packed together and vibrating very slowly. If we heat the metal, the temperature rises as the atoms begin to vibrate more rapidly. If more heat is supplied, the vibrations become more and more violent until a temperature is reached where the atoms are moving so vigorously that they 'escape' from the solid surface. This causes the solid to lose its definite shape and become a liquid. The temperature of the metal remains steady at the melting point until the applied heat has caused all the solid to melt.

In the liquid state, the atoms are not packed in a tight, orderly way. Because they are free to move about randomly throughout the volume of the liquid, heating causes the atoms to rush about more and more rapidly and randomly within the surface of the liquid. At a second constant temperature, the boiling point, the atoms are moving so quickly that they begin to escape from the liquid surface and spread into the surrounding air. The temperature remains steady throughout the change of state but, thereafter, heating results in a rise in temperature due to the increased velocity of movement of the atoms in the gaseous state.

We can see that the heating of a solid, liquid or gas results in expansion because the particles in the body require more and more space for vibration and free movement.

It is important to remember that the atoms and molecules in a liquid are not all moving at the same velocity at any instant. For example, the collision of a pair of moving atoms may result in one being temporarily stationary while the other moves off at a velocity equal to the combined original velocities of the two atoms. Such effects can be clearly demonstrated with the balls on a billiard table. Consequently, since collisions between particles occur continuously, we should think of the temperature of a liquid (solid or gas) as a measure of the mean velocity of the particles where as many particles will have a velocity higher than the mean as lower.

This explains why changes of state are not instantaneous ; the faster particles escape first, leaving the slower particles to be accelerated to the critical velocity for escape by the heat supplied. It also explains the common observation that water evaporates from soils, lakes and even drying washing, at temperatures far below the boiling point of water (100°C). Here again, the very few molecules which are moving at the 'escape velocity' evaporate first. This evaporation causes a loss of kinetic energy to the gas and, therefore, cools the remaining liquid water. However, as long as the water is warmed back to its original temperature by the air or by absorption of solar radiation, there will be a continuous loss of the most energetic molecules. The time taken for complete evaporation of a volume of water at a temperature below its boiling point, will, therefore, depend upon the air temperature, wind speed, and the supply of solar radiation.

When some solids are heated, their atoms or molecules pass directly into the gas state without passing through the liquid state. This change of state, which is relatively uncommon, is called sublimation. The process can be clearly demonstrated by the sublimation of iodine ; on heating, the black crystals of iodine release dense purple fumes of iodine vapour. Similarly, snow can change directly to water vapour under intense solar radiation, for example into the dry air of the Arctic.

The slow loss of molecules from a solid directly to the gas state at low temperatures can be very useful in pest control. For example, the release of molecules from a solid insecticide in a confined space can give protection from insect damage to stored materials such as agricultural products (e.g. grain). Another familiar example is the use of naphthalene (in 'mothballs') to protect stored cloth from the clothes moth.

The law of conservation of matter (or mass)

This law First put forward by the French scientist Antoine Lavoisier in 1774 states: *In a chemical reaction the total mass of the products equals the total mass of the reactants.* Lavoisier's great contributions to chemistry are due in large part to his careful quantitative experiments, i.e., he carefully weighted and measured things. Consequently he performed a number of experiments to verify the law. Although they were as accurate as Lavoisier could make them they would be considered very inadequate by present standards, and it was not until a century later that a series of really accurate experiments was performed by Landolt. His apparatus consisted of an H-tube, as shown in Fig. 5.1. The reagents were placed in different limbs of the tube and, after sealing each limb to ensure that nothing could enter or leave the apparatus, it was weighed. After tilting the tube to permit chemical reaction the apparatus was reweighed. Using a very accurate chemical balance, capable of accuracy to one part in ten million, Landolt found no change in mass.

Landolt was careful to choose reactions with which there was no evolution of gas and little evolution of heat. Such reactions occur on mixing solutions of silver nitrate and sodium chloride, or of sodium sulphate and barium chloride.

Silver nitrate + Sodium chloride → Silver chloride + Sodium nitrate
Barium chloride + Sodium sulphate →Barium sulphate + Sodium chloride.

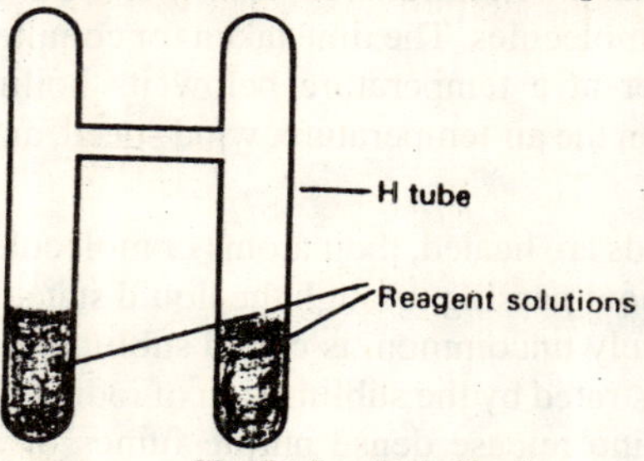

Fig. 5.1 Landolt's H-tube

The choice of such reactions was necessary, because a heavy apparatus would have been required to withstand the pressure of an evolved gas. Further, heat causes the evaporation of a thin film of moisture normally present on the surface of the glass. This means that if the apparatus is weighed before it has time to regain this moisture there is an apparent loss of mass. Because there is no change in mass in the comparatively small number of reactions investigated, it is assumed that the law holds true for all reactions.

The law of constant composition

This law states : *No matter how a given chemical compound is prepared, it always contains the same elements in the same proportions by mass.* It was first stated by Proust in 1799, but was assumed by Lavoisier and other scientists before that date. In common with the Law of Conservation of Mass, it cannot possibly be proved in every case, but the truth of it is accepted on the basis of those compounds which have been investigated.

Law of multiple proportions

This law, which was probably deduced by Dalton and only later subjected to experimental proof, states :

If two elements combine together to form more than one compound, then the masses of one of these elements combining with a fixed mass of the other are in a simple ratio to one another.

For example, carbon forms two common compounds with oxygen : carbon monoxide and carbon dioxide.

With carbon monoxide, 1.33 g of oxygen are combined with 1 g of carbon. With carbon dioxide, 2.67 g of oxygen are combined with 1 g of carbon. That is, the masses of oxygen combining with a fixed mass of carbon are in the ratio 1:2. To verify the law in the school laboratory, examples that use solid substances are more suitable e.g. the oxides of lead.

Example

A metal forms two chlorides containing respectively 65.6 and 55.9 per cent of chlorine. Show that these figures are in accordance with the Law of Multiple Proportions.

The compositions of the chlorides are:

	% chlorine	% metal
Chloride *A*	65.6	34.4
Chloride *B*	55.9	44.1

In chloride *A*,

65.6 g of chlorine combine with 34.4 g of metal.

1 g of chlorine combines with $\frac{34.4}{65.6}$g of metal

$= 0.525$ g of metal

In chloride *B*,

55.9 g of chlorine combine with 44.1 g of metal.

1 g of chlorine combines with $\frac{44.1}{55.9}$g of metal.

$= 0.789$ g of metal

i.e. the masses of metal combining with a fixed mass of chlorine are in the ratio 0.789 : 0.525 = 3:2. These figures are in accordance with the Law of Multiple Proportions.

It should be noted that it is immaterial which element, or what value for its fixed mass, is chosen. If with this same problem, 34.4 g of metal is taken as the fixed mass, we get :

In chloride *A*,

34.4 g of metal combine with 65.6 g of chlorine.

In chloride *B*,

44.1 g of metal combine with 55.9 g of chlorine.

34.4 of metal combine with $\frac{55.9 \times 34.4}{44.1}$ *g* of chlorine

$= 43.6$ g of chlorine

i.e. the masses of chlorine combining with a fixed mass of metal are in the ratio 65.6 : 43.6 = 3 : 2.

TEMPERATURE, TEMPERATURE SCALES AND THERMOMETERS

Temperature is a very important factor in all branches of science. To give a few relevant examples–air and soil temperatures control the growth of plants ; the body temperatures of livestock are useful in the diagnosis of disease ; the burning temperature of a gas mixture determines its usefulness in cutting and welding metals.

In order to measure and compare the temperatures of soils, animals or flames, we require accurate thermometers calibrated over the appropriate range of temperature. The standard temperature scale for most scientific work is the Centigrade or Celsius scale where the melting point of pure ice and the boiling point of pure water (both constant temperatures) are defined as 0°C and 100°C respectively at 1 atm air pressure. Thermometers can be calibrated for use within this range using these fixed points ; for measurements below 0°C or above 100°C, additional changes of state may be used to extend the range of centigrade thermometers.

After the development of the centigrade scale, it was discovered that the lowest possible temperature was – 273°C, i.e. 273 degrees below the melting point of ice. This led to the development of an absolute scale of temperature which retained the same size of units as the centigrade scale but which had its zero point at – 273°C. The scales are, therefore, related as follows :

°A	0	1	2	----	272	273	274	----	372	373	----	
°C	— 273	— 272	— 271	----	– 1	0	1	----	99	100	----	etc.

Although not convenient for scientific work, the Fahrenheit scale is still generally used in everyday life. For example, to most people the air temperature on a hot day will be between 80 and 100° F rather than 27-38°C. When devising his scale in 1714, Fahrenheit chose as his zero point the coldest temperature he could obtain (by mixing equal quantities of snow and ammonium chloride). His 100°F was the temperature of human blood, now known to be 98.4°F. This resulted in temperatures of 32°F for the melting point of water and 212°F for the boiling point. Since one degree Fahrenheit is equal in magnitude to $\frac{5}{9}$ of a degree centigrade, the three scales are related as follows :

$$^\circ C = ^\circ A - 273$$

$$^\circ C = (^\circ F - 32) \times \frac{5}{9}$$

$$^\circ F = \left(^\circ C \times \frac{9}{5}\right) + 32.$$

Since liquids expand on heating, the volume of a fixed mass of liquid can be used as a convenient index of temperature. This is the basis of the familiar glass/mercury thermometer in which variations in temperature can be measured by the expansion and contraction of the liquid metal along a glass tube of uniform internal bore.

Mercury has a unique combination of physical and chemical properties which make it an outstandingly good liquid for general purpose thermometers. As well as ease of visibility and a high thermal conductivity, these properties include :

(*i*) *High Density* Because of its high density (13.6 g ml^{-1} compared with 1 g ml^{1} for water) mercury has a small (but regular) increase in volume per unit increase in temperature. Consequently, glass/mercury thermometers can be compact and portable.

(*ii*) *Liquid Range* Mercury exists as a liquid between its melting point, – 39°C, and its boiling point, 357°C. This liquid range conveniently covers the majority of temperatures encountered in scientific work.

(*iii*) *Surface Properties* The combination of strong forces between the atoms of liquid mercury and very weak attraction between the atoms and glass surfaces ensures that the liquid column does not break as it advances and retreats and does not leave a 'tail' of mercury to obscure temperature readings.

Glass/mercury thermometers cannot be used at extremes of temperature beyond the liquid range of mercury or in field work where the glass is liable to be broken. Under these circumstances it is convenient to use thermocouples or thermistors, whose electrical resistances vary with temperature. When attached to a suitable power source and recorder, these sensors can give continuous records of temperature over prolonged periods.

CHAPTER 6

The Gaseous State

The main differences between the three states of matter were outlined in chapter 5. In this chapter we are concerned with the laws that describe the behaviour of gases.

THE DIFFUSION OF GASES

If a girl using perfume enters a room the smell is rapidly detected all over the room. Less pleasantly, if some hydrogen sulphide is prepared in a laboratory its smell is immediately noticed by everyone present. In both cases the rapid spread of the "smell" is a rapid spread of molecules by a process known as *diffusion.*

Graham's law of diffusion

These experiments indicate that the rate of diffusion of a gas depends on its density, i.e. gases of low density diffuse more rapidly than gases of high density. Thomas Graham investigated this relationship and formulated his *Law of Diffusion :*

At a constant temperature and pressure the rate of diffusion of a gas is inversely proportional to the square root of its density.

Using symbols, this becomes :

$$R \;\alpha\; \frac{1}{\sqrt{d}} \text{ or } R = \frac{k}{\sqrt{d}}$$

(where R = rate of diffusion in $Cm^3\ s^{-1}$ or other suitable units, and d = density). If the rates of diffusion of two gases, A and B, are compared, it follows that :

$$\frac{R_A}{R_B} = \frac{\sqrt{d_B}}{\sqrt{d_A}}$$

Alternatively, the times taken for equal volumes of two different gases to diffuse, under similar conditions, may be compared. Then :

$$\frac{T_A}{T_B} = \frac{\sqrt{d_A}}{\sqrt{d_B}}$$

Molecular masses of gases are directly proportional to their densities. Therefore :

$$\frac{R_A}{R_B} = \frac{\sqrt{M_B}}{\sqrt{M_A}}$$

i.e. Graham's Law affords a means for determining molecular masses.

BOYLE'S LAW

By considering gases to be composed of small particles in constant motion, it is also possible to explain the variation of the volume of a gas with pressure. Suppose that a certain mass of gas enclosed in a container of volume V exerts a pressure P. This pressure is caused by the particles which make up the gas hitting the walls of the container. Provided the temperature of the gas does not alter, the number of collisions with the wall in a particular length of time remains constant.

Now suppose that the same mass of gas is put into a container only half as big, i.e. of volume $\frac{V}{2}$. The number of particles in a particular volume will then be twice as great. This means that collisions with the walls will be twice as frequent and the pressure doubled, i.e. $2P$.

In other words, if the volume of a fixed mass of gas is halved, its pressure is doubled. This was first realised by Robert Boyle in 1662, and is summarised in Boyle's Law:

The pressure of a fixed mass of gas at constant temperature is inversely proportional to its volume.

i.e. Pressure × Volume = Constant

If V_1 and V_2 are the different volumes of a fixed mass of gas at pressures P_1 and P_2, then

$$P_1 V_1 = P_2 V_2$$

CHARLES' LAW

To explain the expansion of gases when they are heated, consider a fixed mass of gas. When the temperature of the gas is raised more energy is supplied to the particles, and they therefore move more rapidly. This means that particles will hit the walls of a container more frequently and, unless the volume is increased, this will result in an increase in pressure.

Charles observed that gases maintained at constant pressure expanded equally for equal rises in temperature. His law states :

A fixed mass of gas, at constant pressure, expands $\frac{1}{273}$ of its volume at 0°C, for each degree Centigrade rise in the temperature, i.e. a gas with a volume of 273 cm^3 at 0°C would expand by 1 cm^3 for each degree rise in temperature and would contract 1 cm^3 for each degree fall in temperature. Fig. 6.1

Theoretically gases would have a zero volume at – 273 °C but in practice they all liquefy above this temperature. This temperature of – 273 °C, at which a gas that did not liquefy would have no volume, is the lowest temperature possible. It is therefore known as the *absolute zero. The*

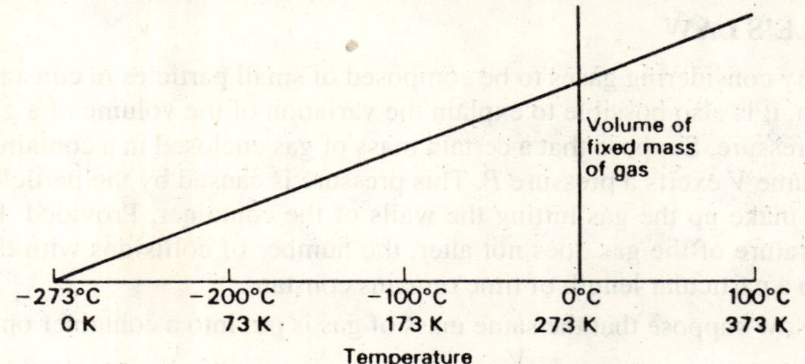

Fig. 6.1. Graph illustrating Charles Law

absolute scale of temperature is obtained merely by adding 273° to the Centigrade temperatures, i.e.

$$x°C = (x+273)K$$

Thus – 273 °C becomes 0 K, 0 °C becomes 273 K, etc. (K is the symbol for the *Kelvin,* the temperature unit on the absolute scale.)

Charles' Law may therefore be stated more conveniently for our purpose:

The volume of a fixed mass of gas, at constant pressure, is directly proportional to its absolute temperature.

i.e. $$\frac{\text{Volume}}{\text{Temperature in K}} = \text{Constant}$$

If V_1 and V_2 are the different volumes of a fixed mass of gas at absolute temperatures T_1 and T_2, then :

$$\frac{V_1}{T_1} = \frac{V_2}{T_2}$$

THE GENERAL GAS EQUATION

By combining the Laws of Boyle and Charles, an expression relating the temperature, pressure and volume of a fixed mass of gas is obtained :

$$\frac{P_1 V_1}{T_1} = \frac{P_2 V_2}{T_2}$$

This is a very important equation. It means, for example, that if the volume of a given mass of gas under one set of conditions is known its volume under a different set of conditions may be calculated. ***Always remember that the temperature used must be on the Absolute scale.***

Normal (or standard) temperature and pressure

When stating the volume or density of a gas it is usual to give the temperature and pressure at which it was fund. This is necessary, as these quantities vary considerably under different conditions.

For comparison purposes, standard values of temperature and pressure, namely 0°C (i.e. 273 K) and 760 mm mercury pressure, have been chosen. These values are known as Normal (or Standard) Temperature and Pressure (abbreviated to n.t.p. or s.t.p.).

Problems on the gas laws

1. *A gas occupies 420 ml at 17 ° C and 770 mm pressure. What is its volume at S.T.P. ?*

$P_1 = 770$ mm $P_2 = 760$ mm

$V_1 = 420 ml$ $V_2 = x\ ml$

$T_1 = 17°C = 290$ K $T_2 = 0°C = 273$ K

$$\frac{P_1 V_1}{T_1} = \frac{P_2 V_2}{T_2}$$

$$\therefore \frac{770 \times 420}{290} = \frac{760 \times x}{273}$$

$$x = \frac{770 \times 420 \times 273}{290 \times 760}\ ml$$

$$= 400\ ml$$

Volume of gas at s.t.p = 400 ml

2. *A gas occupies 36 ml at 27 °C and 750 mm pressure. At what temperature will the same mass of gas occupy 72 ml at 1000 mm pressure ?*

$P_1 = 750$ mm $P_2 = 1000$ mm

$V_1 = 36$ ml $V_2 = 72$ ml

$T_1 = 27°C = 300K$ $T_2 = x$ K

$$\frac{P_1 V_1}{T_1} = \frac{P_2 V_2}{T_2}$$

$$\therefore \quad \frac{750 \times 36}{300} = \frac{1000 \times 72}{x}$$

$$x = \frac{1000 \times 72 \times 300}{750 \times 36} = 800 \text{ K}$$

Temperature is 800 K or (800-273) °C = 527°C

GAY-LUSSAC'S LAW OF COMBINING VOLUMES

It was known during the time of Joseph Priestley that when hydrogen burns it combines with approximately half its volume of oxygen. Henry Cavendish showed that the volumes of the combining gases in this reaction were almost exactly in the ratio 2:1. The French scientist Gay-Lussac was, however, the first to investigate systematically the combining volumes of gases. By an experiment on the reaction of hydrogen with oxygen, he showed that 200 ml of hydrogen combine with 100 of oxygen to give 200 ml of steam, or, in other words, that :

2 vols. of hydrogen + 1 vol of oxygen → 2 vols. of steam

By extending his work to other gaseous combinations he showed that :

1 vol. of hydrogen + 1 vol. of chlorine → 2 vols. of hydrogen chloride

2 vols. of carbon monoxide + 1 vol. of oxygen → 2 vols. of carbon dioxide

1 vol. of hydrogen chloride + 1 vol. of ammonia → solid ammonium chloride

1 vol. of nitrogen + 3 vols. of hydrogen → 2 vols. of ammonia

In all the cases he investigated he noticed the simple relationship existing between the volumes of the gases involved, and in 1808 he advanced his now famous law :

When gases combine, they do so in volumes which bear a simple ratio to one another, and to the volume of the product if it is gaseous, provided all the volumes are measured at the same temperature and pressure. As volumes vary with temperature and pressure, the reason for the last part of his law should be obvious.

The law *only applies to gases.* If hydrogen and oxygen were combined under such conditions that water, and not steam, was formed the volume of the water would be negligible and would bear no simple relationship to the volume of the gases. Such is the case, also, with the formation of solid ammonium chloride and numerous other reactions, e.g. :

Solid carbon + 1 vol. of oxygen →1 vol. of carbon dioxide

Solid sulphur + 1 vol. of oxygen → 1 vol. of sulphur dioxide

It is difficult to see at first why 2 volumes of hydrogen and 1 volume of oxygen should give only *two* volumes of steam. This is because steam is $\frac{3}{2}$ times denser than the mixture of hydrogen and oxygen, but this whole question should become clearer later in the chapter.

A certain similarity exists between Gay-Lussac's Law and one of the main points in Dalton's theory. Gay-Lussac stated that :

When volumes of gases combine they do so in numbers which bear a simple ratio to one another.

Dalton's theory stated that :

When atoms combine they do so in numbers which bear a simple ratio to one another. This similarity led J. J. Berzelius, a Swede, to suggest that equal volumes of all gases at the same temperature and pressure contain the same number of atoms or "compound atoms" Unfortunately, deductions made from this suggestion disagreed with Dalton's theory. For example :

1 vol. of hydrogen + 1 vol. of chlorine → 2 vols. of hydrogen chloride.

If Berzelius' suggestion is correct, and 1 volume of hydrogen contains *n* atoms, then 1 volume of chlorine will also contain *n* atoms, and 2 volume of hydrogen chloride will contain 2*n* "compound atoms", i.e.

n atoms of hydrogen + *n* atoms of chlorine → 2*n* "compound atoms" of hydrogen chloride, or

1 atom of hydrogen + 1 atom of chlorine → 2 "compound atoms" of hydrogen chloride.

Therefore 1 "compound atom" of hydrogen chloride must contain $\frac{1}{2}$ an atom of hydrogen and $\frac{1}{2}$ an atom of chlorine. Accepting the fact that the atom is indivisible, this statement is absurd.

AVOGADRO'S LAW

To Amedeo Avogadro the Italian, we owe the conception of the molecule, and the abandoning of the term 'compound atom'. The term 'molecule' applies to elements and compounds, and is defined as *the smallest part of an element or compound that can exist alone.* This means that elements do not necessarily exist as single atoms, but may exist as groups of two or more atoms. The number of atoms in a molecule of an element is referred to as its *atomicity.* Thus helium, which contains one atom per molecule, is monatomic ; hydrogen, which contains two atoms per molecule, is diatomic and so on. In 1811 Avogadro formulated his law :

Equal volumes of all gases at the same temperature and pressure contain the same number of molecules.

Applying this relationship to the combination between hydrogen and chlorine, we get :

1 vol. of hydrogen + 1 vol. of chlorine → vol. of chlorine → 2 vols. of hydrogen chloride

n molecules of hydrogen +*n* molecules of chlorine → 2*n* molecules of hydrogen chloride

1 molecule of hydrogen + 1 molecule of chlorine → 2 molecules of hydrogen chloride.

This means that if the molecules of hydrogen and chlorine contain an even number of atoms this deduction is not in conflict with Dalton's theory.

There is much evidence for believing that the hydrogen chloride molecule does in fact contain only one atom of hydrogen and one atom of chlorine combined together. Where the hydrogen is concerned, for example, hydrogen chloride gives rise to only one sodium compound by substitution of hydrogen with sodium, namely sodium chloride, and that compound contains no hydrogen. Similar compounds, such as sulphuric acid, which contains two hydrogen atoms, form two sodium derivatives : sodium sulphate, $Na_2\ SO_4$, and sodium hydrogen sulphate, $NaHSO_4$. Since hydrogen chloride contains only one hydrogen atom, the hydrogen molecule must be diatomic. (Further evidence in support of the diatomicity of hydrogen is beyond the scope of this book.) By similar reasoning the chlorine molecule can be shown to be diatomic.

The reaction between hydrogen and chlorine now becomes clear if, instead of using the more conventional equation, we represent it in this way :

$$HH + ClCl \rightarrow HCl + HCl$$

i.e. one molecule of hydrogen containing two atoms combines with one molecule of chlorine containing two atoms, to give two molecules of hydrogen chloride each containing one atom of hydrogen combined with one atom of chlorine.

It is important to realise exactly the difference between an *atom* of an element and a *molecule* of an element. An atom is the smallest particle of an element that can take part in a chemical change. Notice that single atoms of hydrogen and chlorine are present in a molecule of hydrogen chloride. Therefore, in accordance with the definition, these atoms can be considered to have taken part in a chemical change.

A molecule of an element is the smallest part of it that can exist alone. Therefore the hydrogen and chlorine, before they have taken part in a chemical reaction, are represented as aggregates of two atoms, i.e. as molecules, in which form those elements normally exist.

The Avogadro constant

In chapter 2, where the *mole* was discussed, we explained that the number of particles in a mole of a substance is 6.02×10^{23}, and that this number is known as *Avogadro's constant.*

It is possible therefore to restate *Avogadro's Law,* which refers to gases only, as : *Equal volumes of all gases at the same temperature and pressure contain the same number of moles.*

50 cm^3 propane + 250 cm^3 oxygen → 150 cm^3 carbon dioxide, i.e. 250 cm^3 oxygen are consumed and 150 cm^3 carbon dioxide are formed.

THE LIQUEFACTION OF GASES

Chapter law appears to describe the volume changes of gases at all temperatures from OK upwards. However, as a gas is called, its molecules move more and more slowly until they finally condense to give a liquid at a temperature normally well above OK. Consequently, Charles' Law can be applied to a substance only at temperatures above its boiling point (at the fixed pressure used).

Increase of pressure also causes gases to condense to liquids by forcing the molecules together. For example, we can keep liquid propane under pressure in a cylinder at laboratory temperatures (15–25°C) which are far higher than the boiling point of propane (–42°C). Thus Boyle's Law is applicable only over a limited range of pressures.

The liquefaction of gases by pressure is an important feature of the petroleum fuel industry. In many parts of the world the hydrocarbon gases, methane (b. pt. 164°C), ethane (–89°C), propane (–42°C) and butane (–0.5°C), occur in huge quantities either in natural gas fields or in oil wells. These fuels must be liquefied for efficient transport and distribution but this cannot be achieved economically by costly, long-term refrigeration. Instead, the gases are liquefied by pressure and stored in strong pressurized containers for example, in 20 kg laboratory cylinders of liquid propane. No energy is needed to maintain the fuel as a liquid and when a flow of fuel gas is required, it can be supplied by opening the valve in the cylinder thus reducing the pressure and permitting the liquid to vaporize.

METHODS OF EXPRESSING THE COMPOSITION OF GAS MIXTURES

The composition of gas mixture is commonly expressed in terms of the *percentage* of the total *volume* occupied by each gas. Thus the composition of dry air is often quoted as 78% nitrogen (N_2), 21% oxygen (O_2), 0.93% argon (Ar), 0.03% carbon dioxide (CO_2) (variable), *by volume,* the remaining volume being occupied by small quantities of the other inert gases (He, Ne, Kr), nitrogen oxides and methane. Minor constituents of air (e.g. pollutants like SO_2 and nitrogen oxides) are normally expressed as p.p.m. (volumes per million volumes – e.g. mL per Litre).

The volume percentage composition of a gas mixture is very useful because it also gives information on the ratio of the number of moles of each constituent gas. This is because, by *Avogadro's Principle* :

"Equal volumes of different gases contain the same number of moles (or molecules) if they are at the same temperature and pressure. For example, it has been found that 1 mole of any gas at STP has a volume of 22.41."

From this we can calculate that a 22.41 sample of air at STP contains 0.78 mole of N_2, 0.21 mole of O_2 and 0.0093 mole of Ar.

Avogadro's Principle can also be written as a form of the General Gas Equation :

$$\frac{pV}{T} = nR$$

or

$$V = \frac{nRT}{p},$$

where n is the number of moles in the gas sample and R is constant, (the Gas Constant) i.e. the volume of a gas sample at constant pressure and temperature is proportional to the number of moles of gas in the sample. This form allows us to explain another widely used method for expressing the composition of a gas mixture–*partial pressures.* For example, in a given volume of air (V) contained in a bottle, the total number of moles present (n) is equal to the sum of the number of moles of the constituent gases :

$$n = n_{N_2} + n_{O_2} + n_{Ar} \text{ etc.}$$

For a given temperature, T, we can multiply each side of this equation by the constant $\frac{RT}{V}$, giving :

$$\left(\frac{RT}{V}\right)n = \left(\frac{RT}{V}\right)n_{N_2} + \left(\frac{RT}{V}\right)n_{O_2} + \left(\frac{RT}{V}\right)n_{Ar}$$

By the G.G.E., the left-hand side of this equation, $\frac{RT}{V}n$, is now equal to p, the pressure exerted by the volume of air against atmospheric pressure (1 atm at STP – note that if a volume of gas did not exert a pressure against atmospheric pressure, it would collapse). Similarly, $\frac{RT}{V}n_{N_2}$ is equal to the *partial pressure* of nitrogen (p_{N_2}), that is, the pressure it would exert if it *alone* filled the bottle and the other constituents were absent. Similarly $\frac{RT}{V}n_{O_2} = p_{O_2}$, the partial pressure of oxygen, i.e.

$$p = p_{N_2} + p_{O_2} + p_{Ar}.$$

We have already seen that 1 'mole' of air occupies 22.4 1, exerts a pressure of 1 atm and contains 0.78 mole of N_2, 0.21 mole of O_2 and 0.0093 mole or Ar. Consequently, since

$$p = \frac{RT}{V}n = 1, \frac{RT}{V} = 1$$

and, therefore, the partial pressures of nitrogen, oxygen and argon in dry air at STP are

$$p_{N_2} = 1 \times 0.78 = 0.78 \text{ atm}$$

$$p_{O_2} = 1 \times 0.21 = 0.21 \text{ atm}$$

$$p_{Ar} = 1 \times 0.0093 = 0.0093 \text{ atm}.$$

Note that the ratio of partial pressures of the different components is the same as the ratio of the number of moles of the components.

So far we have discussed the composition of dry air only. However, since the moisture content of the air is one of the most important environmental factors controlling plant growth, it is necessary to understand the conventional methods of expressing water vapour content.

The simplest expression employed is the *Specific Humidity* which is the mass of water vapour contained in unit mass of *moist* air (g g^{-1} or kg^{-1}). The maximum specific humidity of air, at which it is saturated with moisture, varies widely with temperature (Table 6.1) and air pressure.

Table 6.1. Quantities of water vapour required to saturate air at 1 atm pressure and various temperatures (g kg^{-1} of moist air).

Temperature (°C)	*g kg^{-1}*
0	3.8
5	5.4
10	7.7
15	10.7
20	14.7
25	20.0
30	26.9
35	35.8
40	47.3

Because of the great variability of air moisture content, another expression *Relative humidity* is more commonly used. The relative humidity. (RH) of air is calculated using the following formula :

$$RH(\%) = \frac{\text{water vapour content of the air (g kg}^{-1}\text{ air)}}{\text{water vapour content of saturated air at the same temperature and pressure (g kg}^{-1}\text{ air)}} \times \frac{100}{1}.$$

Consequently, the *RH* of saturated air is always 100%, whatever the pressure or temperature, and any water vapour content lower than the saturation value gives an *RH* value less than 100%. For example, using Table 6.1 we can calculate that the *RH* of air containing 5.4 g kg^{-1} water vapour falls from 100% at 5°C, through 50% at 15°C and 27% at 25°C to 15% at 35°C (and 1 atm air pressure).

For reference purposes, some important properties of oxygen and carbon dioxide in aqueous solution are given in Table 6.2.

Table 6.2. Gas solubilities and diffusion coefficients in water.

	10°C	*20°C*	*30°C*
Solubility of O_2 (mole l^{-1})	0.00035	0.00028	0.00024
Solubility of CO_2 (mole l^{-1})	0.0514	0.0365	0.0267

Diffusion coefficient of O_2 in water at 25°C $= 2.9 \times 10^{-5}$ $cm^2\ s^{-1}$

Diffusion coefficient of CO_2 in water at 20°C $= 1.7 \times 10^{-5}$ $cm^2\ s^{-1}$

(The diffusion of O_2 and CO_2 is much faster through air than through water as shown by their D values in air : 1.8×10^{-1} $cm^2\ s^{-1}$ for O_2 and 1.4×10^{-1} $cm^2\ s^{-1}$ for CO_2, both value measured at 0° C).

CHAPTER 7

The Liquid State

Few substances occur as pure liquids in everyday life although mercury in thermometers is a familiar exception. Instead, most liquids are solutions in which one or more substances (solutes) are dissolved in a pure liquid (solvent).

A *solution* is formed when a solute is separated into its constituent particles (molecules, ions, atoms) and dispersed uniformly throughout a solvent.

In this chapter, we shall confine our discussion to aqueous solutions in which the solvent is liquid water. However, both solute and solvent can exist in each of the three states of matter. For example, many alloys consist of one solid metal dissolved in another; similarly, air may be considered to be a solution of oxygen, carbon dioxide and the inert gases in nitrogen as solvent. An understanding of the properties of aqueous solutions is an essential part of all studies of environmental chemistry, because of the central role and aqueous solutions play in environmental process. For example:

(*a*) A large proportion of the Earth's total primary production originates from aqueous solution (marine and freshwater ecosystems).

(*b*) Most biochemical reactions take place in the aqueous environment of plant cell, animal cells and animal body fluids.

(*c*) Pesticides and some nutrients are applied to crops and live-stock as dilute aqueous solutions (e.g. fungicide and herbicide sprays in crop protection; foliar nutrient sprays; treatment of livestock against ticks by spraying or dipping). These agrochemicals can be applied more safely, accurately and uniformly in solution than as dry solids.

(*d*) Other aqueous solutions include the soil solution, which is normally the only source of inorganic nutrients for plant uptake, and irrigation water, whose solute content must be low to prevent the development of salinity in soils.

In this chapter, some of the basic properties of aqueous solutions are discussed.

Solution Concentration Expressions

In the preparation and use of solutions, it is necessary to express the concentration of dissolved solute in an exact way. The simplest form of expression, mass of solute per unit volume of solvent (e.g. grams/litre) is familiar in everyday life. However, in experimental work, where the effects of two or more different solutes are being compared, it is more convenient to use molarity to express solute concentration. Before we can use this expression, it is necessary to define some new terms.

***The Molecular Weight* (*MW*)** of a covalently bonded compound is the sum of the atomic masses of the atoms in one molecules of the compound. Thus for urea, CON_2H_4, which has 1 carbon, 1 oxygen, 2 nitrogen and 4 hydrogen atoms in each molecules.

$$MW = 1(12) + 1(16) + 2(14) + 4(1) = 60.$$

Although there are no molecules in ionic compounds the term molecular weight is still used. The MW of an ionic compound is the sum of the atomic masses of the atoms in the *formula* (which expresses the simplest ratio of ions in the compound, i.e. KCl, not K_2Cl_2 or K_3Cl_3, etc). Therefore, for potassium chloride, KCl,

$$MW = 1(39) + 1(35.5) = 74.5$$ (occasionally this is called the formula weight).

Note that the strictly correct term Molecular Mass is never used.

The **Gram Molecular Weight** (*GMW*) of a compound is its molecular weight expressed in grams (rather than daltons) and is the mass of 1 **Mole** of the compound. The GMW of urea and potassium chloride are, therefore, 60 g and 74.5 g respectively.

A *l* **Molar** solution of a substance contains 1 gram molecular weight (or 1 mole) of the substance dissolved in 1 litre of solution. For example a 1 M solution of urea contains 60 g urea per litre and a 2 M solution of KCl contains 149 g Kcl per litre.

The molarity expression of concentration is particularly useful because it stresses the *number* of molecules (or ions) of the solute per litre. For example, since the mass of each molecule of urea is 60 atomic mass units, *amu* ($1 \text{ amu} = 1.66 \times 10^{-24}$g) and the mass of 1 mole is 60 g, then the number of molecules in a mole is:

$$\frac{60}{1.66 \times 10^{-24}} = 6.02 \times 10^{23}$$

6.02×10^{23} is called the *Avogadro Number*. We can, therefore, conclude that all 1 molar solutions contain the same number of molecules per litre, whatever the solute.

This can be extended to ionic solutes. For example, 1 mole of KCl contains 1 Avogadro Number of K^+ ions *and* 1 Avogadro Number of Cl^- ions. The total number of particles per mole will, therefore, be 12.04×10^{23}. In 1 mole of potassium sulphate, K_2SO_4, there are 2 Avogadro numbers of K^+ ions and 1 Avogadro number of SO_4^{2-} ions giving a total of 18.06×10^{21} ions.

The value of the use of moles rather than masses can be illustrated by a simple experiment in plant nutrition. For example, if we have pure samples of the following nitrogen fertilizer substances:

The value of the use of moles rather than masses can be illustrated by a simple experiment in plant nutrition. For example, if we have pure samples of the following nitrogen fertilizer substances:

Anhydrous ammonia NH_3

Sodium nitrate $NaNO_3$

Urea CON_2H_4,

how can we find out which of these forms of nitrogen is most easily absorbed by the roots of a crop growing in soil? As a first step we might establish the crop in pots, apply equal amounts of the fertilizer compounds (say 10 g) to different pots and then assess the availability of the nitrogen applied by measuring the amount of nitrogen in the leaves of the crop after a given period of growth. However, in applying the same mass of each fertilizer, we are applying different amounts of nitrogen:

Ammonia GMW = 1(14) +(1) = 17 g containing 14 g N
Therefore 10 g contains *8.2 g nitrogen;*

Sodium Nitrate GMW = 1(23) + 1(14) + 3(16) = 85 g containing 14 g N

therefore 10 g contains 1.6 g nitrogen;

Urea GMW = 1(12) + 1(16) + 2(14) + 4(1) = 60 g containing 28 g N

therefore 10 g contains *4.7 g nitrogen.*

Consequently, we can obtain no useful information from this experiment because the amounts of nitrogen applied, as well as its availability, very amongst the experimental treatments. If, instead, we apply 1 mole of ammonia (17 g), 1 mole of sodium nitrate (85 g) or 0.5 mole urea (30 g) to the pots, then the soil receives the same amount of nitrogen (14 g) in each case and differences in leaf N content will reflect differences in the availability of the nutrient.

Other solution concentration terms used widely in chemistry are *normality* (number of gram equivalent weights per litre) which is considered in Chapter 11, and ppm (parts per million, i.e. number of grams per million grams of solution) which is used for very low concentrations of solute, for example for the trace element contents of plants and soils.

SOLUBILITY

There is a limit to the amount of solute which will dissolve in a solvent and when this limit is reached, the solution is said to be saturated with the solute. The concentration of solute in a saturated solution is the *solubility* of the solute in the solvent, normally expressed as gL^{-1} or g 100 mL^{-1}. The solubility of solutes in water varies widely between zero and over a thousand gL^{-1} according to the chemical nature of the solute, as shown for a selection of compounds in Table 7.1. Clearly, the very low solubilities of some pesticides mean that they connot be sprayed as aqueous solutions; in particular, sulphur, DDT and Dieldrin are normally

TABLE 7.1. The water solubility of a selection of substances of importance in agriculture and pollution (most values measured at 0°C or 20°C; compiled from a number of sources).

Solute	Solubility (gL^{-1})
Ammonium nitrate	1183
Potassium chloride	347
Calcium sulphate	2
Cooper sulphate	316
Sulphur	0 (totally insoluble)
Urea	1000
DDT (insecticide)	0 (totally insoluble)
Dieldrin (insecticide)	1.86×10^{-4}
Simazine (herbicide)	$5 < 10^{-3}$

applied as dry powders. On the other hand, the very high solubility of ammonium nitrate can lead to the pollution of drainage water and water-courses by fertilizer nitrate.

The solubility of solutes is governed by three general rules:

(1) The solubility of most compounds in water increases with increasing temperature;

(2) The solubility of most gases in water decreases with increasing temperature;

(3) Substances tend to dissolve in solvents which are chemically similar to them. Thus ionic substances, which are made up of charged ions, dissolve readily in polar solvents, like water, whose molecules carry partial charges due to the polarization of covalent bonds. Such solutes, which are called *hydrophilic* (water loving) are relatively insoluble in non-polar solvents. In contrast, un-polarized molecular solutes (including many organic compounds) have very low solubilities in water but dissolve freely in non-polar solvents like benzene, paraffin and petrol. Both solutes and solvents are therefore *hydrophobic* (water hating').

Although most ionic compounds (salts) do dissolve readily in water, there is a group of important inorganic salts which are relatively insoluble, as shown by the following classification:

Soluble Salts (*solubility* 10 g L^{-1})–All nitrates and acetates; most chlorides, bromides, iodides and sulphates; all salts of Group 1 metals (Na, K, etc.) and ammonium.

Insoluble Salt (*solubility* 1 gL^{-1})–Most hydroxides, carbonates, phosphates and sulphides (except those of the Group 1 metals and ammonium).

Thus, of the three macronutrients in plant nutrition, N, as nitrate of ammonium ions, and K, as potassium ions, are highly soluble, whereas P, as phosphate, is relatively insoluble.

THE SOLUBILITY PRODUCT

Because hydroxides, carbonates and phosphates are involved in many aspects of soil and water chemistry, it is necessary to have a more precise way of expressing and predicting the water solubility of these sparingly soluble salts. This may be achieved using solubility products.

The **Solubility Product Law** states that in all saturated solutions of a given sparingly soluble salt, the product of the concentrations (moles/litre) of the component ions is the same. This constant product is called the solubility product, and values for some selected salts are shown in Table 7.2. For example, in all saturated solutions of silver chloride, AgCl,

$$\text{Solubility Product } (K_{AgCl}) = [Ag^+] \times [Cl^-] = 1.6 \times 10^{-10},$$

where *square brackets indicate molar concentrations*. Similarly, for all saturated solutions of aluminium hydroxide, $Al(OH)_3$.

$$K_{Al(OH)_3} = [Al^{3+}] \times [OH^-] \times [OH^-] \times [OH^-]$$

$$= [Al^{3+}]\,[OH^-]^3 = 1 \times 10^{-33}.$$

Note that each ion in the formula is involved in the product.

The lower the value of K, the lower is the solubility of the salt. Thus hydroxyapatite, which is a major phosphate mineral in rocks, is effectively insoluble in water (Table 7.2).

Although the solubility product law refers to solutions of one solute only, under certain conditions we can use the law to predict the results of adding a second solute. For example, if the second salt does not contain

TABLE 7.2. The solubility products of some sparingly soluble salts, mainly at 18-25°C.

$CaSO_4 \cdot 2H_2O$	2.4×10^{-5}
$Ca(OH)_2$	8×10^{-6}
$CaHPO_4$	1.3×10^{-8}
$CaCO_3$	4.8×10^{-9}
AgCl	1.6×10^{-10}
$Fe(OH)_2$	1×10^{-15}
$Al(OH)_2(H_2PO_4)$*	2.8×10^{-29}
$Al(OH)_3$	1×10^{-33}
$Fe(OH)_3$	1×10^{-38}
$Ca_{10}(PO_4)_6(OH)_2$†	1.5×10^{-112}

*Variscite–an important phosphate mineral in soils.

†Hydroxyapatite–one of the minerals in rock phosphate.

an ion common to the first (e.g. $CaCO_3$ and AgCl) then each salt will tend to dissolve as if the other were not present.

In contrast, if the two salts do have a common ion, then the addition of the second salt may influence the solubility of the first. As a simple example, if we add some solid KOH to a saturated solution of $Al(OH)_3$ then the hydroxide ion concentration in the solution $[OH^-]$, rises due to the high solubility of the KOH. However, according to the solubility product law, $[Al^{3+}]\,[OH^-]^3$ must remain constant and therefore $[Al^{3+}]$ must decrease, i.e. Al^{3+} ions come out of solution, giving a precipitate of solid $Al(OH)_3$. Thus the addition of KOH causes a lowering of the solubility of aluminium ions.

This phenomenon occurs in soils where Al^{3+} ions, which are highly toxic to most plant species, become less soluble as the hydroxide ion concentration of the soil solution rises. Thus, only those species whose roots are tolerant of Al^{3+} ions can grow successfully in soils with low hydroxide ion concentration (pH 5). On the other hand, crop plants, which are generally sensitive to Al^{3+}ions, are grown in agricultural soils which are maintained at moderate hydroxide ion concentrations (pH 5.5–7.0) by liming.

Overall, it is important to stress that solubility products can be used in environmental chemistry only with great care. Firstly, since most naturally occurring aqueous solutions contain more than one sparingly soluble salt, common ion effects must be taken into account. Secondly, the presence in solution of certain organic compounds (chelating agents) can result in large increase in the solubility of ionic solutes. For example, the molecules of fulvic acid, a soluble component of soil humus, can 'wrap around' lead ions, thereby rendering them soluble and subject to leaching. This is an important process in the mobilization of heavy metal pollutants. In the same way, the elevated (bleached) E horizons of podzols are caused by enhanced solubility and leaching (cheluviation) of Fe^{3+} and Al^{3+} ions due to polyphenol chelating agents released by heath and coniferous vegetation. The synthetic chelating agent EDTA is used in the chemical analysis of sparingly soluble metal ions.

DIFFUSION AND OSMOSIS

In an aqueous solution, both the water molecules and the atoms, ions or molecules of the solute are in content, rapid and random motion. Consequently, any differences in solute concentration which might occur within the solution are quickly abolished by the random movement of solute particles. This movement is called diffusion.

The process of diffusion can be illustrated as follows. The volume of a beaker is divided into two equal parts (A and B) by a watertight partition, A containing 50 ml of 1 M NaCl solution and B containing 50 ml of 2 M NaCl solution. Because there are twice as many Na^+ and Cl^- ions in B than A, the random movement of the solute results in twice as many Na^+ and Cl^- ions per second colliding with the B side of the partition than the A. Consequently, when the partition is removed, twice as many ions per second pass from B to A than pass from A to B and the difference in concentration decreases rapidly to give 100 ml of 1.5 M NaCl solution. Na^+ and Cl– ions have diffused from a region of high concentration (B) to a region of low concentration (A) by random movement. This movement can be described as diffusion down a concentration gradient.

The diffusion of solute particles from a region of high solute concentration (c_2) to a region of low concentration (c_1) can be described by a form of Fick's First Law:

$$\text{Diffusion Rate} = AD\frac{c_2 - c_1}{l}$$

where A is the cross-sectional area through which diffusion is occurring, l is the distance between c_1 and c_2 and D (the diffusion co-efficient of the solute in water) is the rate of diffusion when

$$\frac{A(c_2 - c_1)}{l} = 1$$

(e.g. a concentration gradient of 1 mole L^{-1} cm^{-1} and a cross-sectional area of 1 cm × 1 cm.)

D values can, therefore, be used as a measure of the relative rates at which different solutes diffuse down a concentration gradient. For example, Table 7.3 gives *D* values for nitrate, phosphate and potassium ions in solution and in moist soil. The values are much lower and, therefore, diffusion is much slower in soil because of the long and 'tortuous' diffusion pathway though the water films round soil particles (since ionic diffusion does not occur through air filled soil pores or through the solid soil particles). The value for phosphate is especially low because phosphate ions tend to precipitate in the soil as they pass along the diffusion path.

TABLE 7.3. Diffusion coefficients for ions in solution and in moist soil at 20-25° C (cm^2 s^{-1}).

	Solution (× 10^{-5})	*Soil* (× 10^{-5})
Nitrate	1.92	0.5
Potassium	1.98	0.01-0.24
Phosphate	0.89	0.0005-0.001

(Note that the diffusion coefficient of an ion in solution is related to its molecular weight; for example, the phosphate ion is both the heaviest and the slowest of the three ions.)

Because of lower densities and higher particle velocities, diffusion rates are much higher in the gas state than in aqueous solution. In contrast, diffusion is extremely slow in solids.

Returning to our beaker, if we replace the partition between A and B with a membrane which is permeable to Na^+ and Cl^- ions, then the equalization of NaCl concentration in A and B occurs by diffusion through the membrane. However if we replace the permeable membrane with a *semi-permeable* membrane, which is permeable to water but not to solutes, then the Na^+ and Cl^- ions cannot diffuse through the membrane from B to A to give equalization of solute concentration. Instead, water, which is at a higher 'concentration' in A, diffuses through the membrane from A to B because of the random movements of its molecules. This movement of water across a semipermeable membrane in response to a difference in solute concentration is called *osmosis*.

Strictly speaking, in osmosis, water moves in response to differences in *osmotic pressure* rather than solute concentration. The osmotic pressure (π) of a solution is directly proportional to the solute concentration (c) and can be calculated using the Van't Hoff equation:

$$\pi = icRT,$$

where R is a constant, T is the absolute temperature and π is expressed in atmospheres or bars (1 atm = 1.013 bars). For molecular solutes, $i = 1$, but for ionic solutes i is the number of ions in the formula (e.g. $i = 2$ for KCl, 3 for K_2SO_4, etc.). Thus, if c is expressed in moles/litre, ic is a measure of the total number of particles per litre which cannot cross a semi-permeable membrane. Osmotic pressure is, therefore, determined by the number of particles, rather than the molar concentration, of solute.

It might be expected that osmotic movement of water from the weaker (A) to the stronger (B) solution would continue until the concentrations of NaCl were equal at 1.5 M. However, as osmosis proceeds, a head of water develops between A and B–

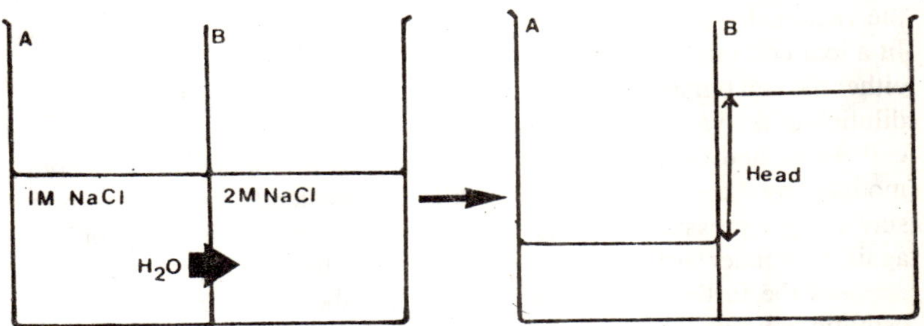

whose hydrostatic pressure opposes further water movement. At equilibrium, the osmotic flow of water ceases because the pressure of the head of water is equal to the difference in osmotic pressure between the solutions in A and B (note that this is the osmotic pressure difference at equilibrium, not at the beginning of the experiment).

A good example of this process is provided by plant cells. In a typical leaf cell, the large central vacuole contains a concentrated solution of organic and inorganic solutes at an osmotic pressure of about 16 bars. The vacuolar contents are separated from the very dilute aqueous solution (osmotic pressure $\simeq$ 0) bathing the outside of the cell, by the tonoplast membrane, the cell cytoplasm and the plasmalemma, which together act as a semi-permeable membrane (Fig 7.1). Consequently, if the leaf is not

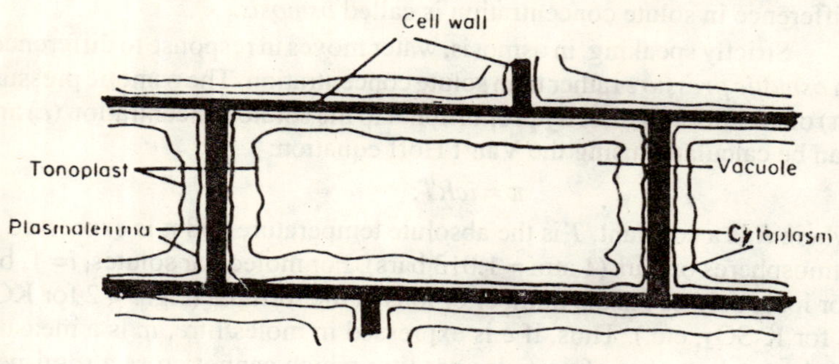

Fig. 7.1. Schematic cross-sectional diagram of a typical mature plant cell showing the 'complex semipermeable membrane' (plasmalemma, cytoplasm and tonoplast) separating water in the cell walls from that in the vacuole. Note that cytoplasmic inclusions (nuclei and organelles) are omitted and that, in a fully turgid cell, the plasmalemma would be pressed against the cell walls.

transpiring rapidly, water tends to move from the external solution into the vacuole by osmosis, and the volume of the vacuole tends to increase. In a leaf cell lacking cell walls, this flow of water would continue until either the cell burst or the difference in osmotic pressure was removed by dilution of the vacuolar sap. However, in a leaf, cell volume is limited by cell walls, and only a relatively small inflow of water can be accommodated by the elasticity of these walls. Consequently, hydrostatic pressure (turgor pressure) develops in the vacuole, pressing the cytoplasm against the inner surface of the cell walls. When the turgor pressure, which opposes the further entry of water, reaches about 12 bars it equals the osmotic pressure difference (reduced to about 12 bars by dilution of the vacuole) which is driving water inwards. Osmotic water flow then ceases and the cell is at maximum turgor pressure.

It may be helpful to think of the plant cell as a balloon, whose skin is the compound semi-permeable membrane of tonoplast, cytoplasm and plasmalemma, being blown up inside a rigid box, the cell walls. When the balloon fills the internal volume of the box, its volume can increase no further; the air pressure in the balloon then increases (development of turgor pressure) until it equals the maximum pressure the 'blower' can exert (the difference in osmotic pressure between vacuole and exterior). Further inflation is then impossible (osmotic inflow of water ceases). Note that without the box, inflation would continue until the balloon burst.

CHAPTER 8

The Solid State

The atoms, ions or molecules in solids are tightly packed together. According to the regularity of this packing, solid substances can be classed as crystalline or amorphous. *Crystalline solids* (Crystals) have regular three-dimensional shapes with flat faces, straight edges, sharp vertices and places of weakness parallel to their faces. This regularity of shape is a direct consequence of the regularity of packing of the particles in the solid.

Amorphous solids do not have regular three-dimensional shapes. Most of the substances studied in chemistry are crystalline although a few, such as proteins and clay-sized minerals, may be amorphous. In this chapter as well investigate some relationships between the crystal structure (arrangement of particles) or solid substances and their bulk properties.

CRYSTAL STRUCTURES

Ionic and metallic substances have close-packed crystal structures, as shown for NaCl in Fig. 8.1 In NaCl the maximum number of negative chloride ions (6) is packed round each positive sodium ion, and the maximum number of soidium ions (6) is packed around each chloride ion. This arrangement is repeated throughout the crystal giving a uniform structure held together by strong ionic bonds. Many ionic compounds, including KCl, have similar cubical crystal structures.

Because of uniform, strong bonding in their crystals, ionic solids tends to be mechanically strong and to have high melting points since a great deal of energy must be supplied to enable ions to escape from the solid surface. For example, NaCl melts at 801°C and KCl at 776°C.

The structures of ionic solids also account for the characteristically regular shapes of their crystals. In our example, NaCl crystals are cubic simply because the arrangement of Na and Cl ions in the crystals is cubic;

the crystals split naturally along planes of weakness between the layers of ions to give regular faces, edges and vertices (Fig. 8.1).

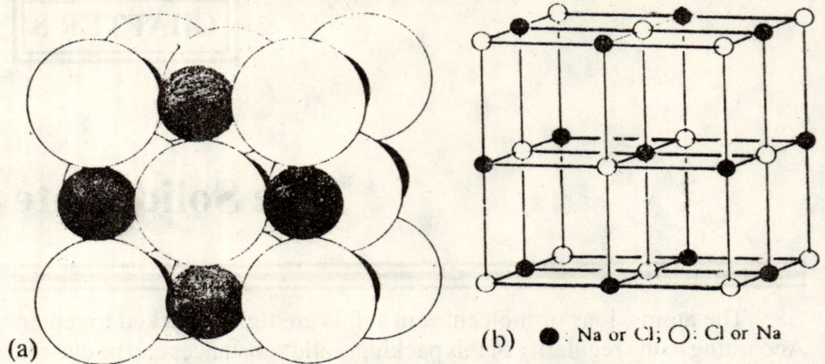

Fig. 8.1 The crystal structure of sodium chloride, NaCl. (a) Showing the close-packed arrangement, where the sodium atoms are represented by the smaller spheres, and (b) stressing the relative positions of the ions in the structure.

Covalently bonded substances are also normally crystalline in the solid state. Their molecules are arranged in a three-dimensional pattern which, as in ionic solids, repeats regularly throughout the crystal. This can be

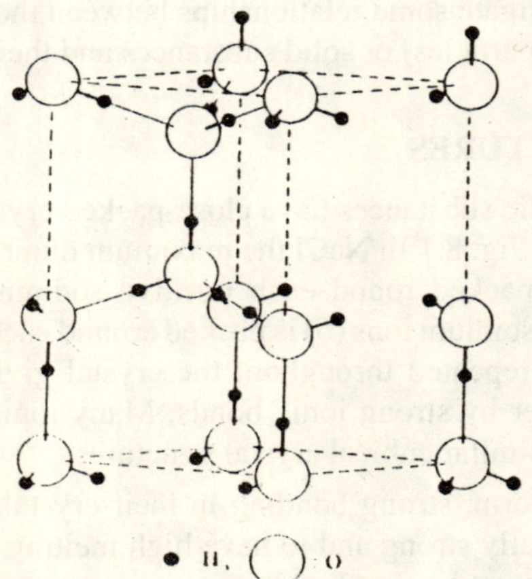

Fig. 8.2 The hexagonal crystal structure of ice showing the regular three-dimensional arrangement of water molecules held together by hydrogen bonds.

illustrated by Fig. 8.2 which shows one of the crystal structures of ice. Because the intramolecular forces (covalent bonds) in their crystals are stronger than the intermolecular forces (van der Walls or Hydrogen bonds), covalent solids tend to be physically weaker than ionic solids and to have lower melting points. For example, carbon dioxide (CO_2) melts at – 56.6°C and water at 0°C.

CRYSTAL STRUCTURES AND MECHANICAL PROPERTIES.

1. The Allotropes of Carbon

One of the best examples of the relationship between crystal structure and bulk mechanical properties is provided by the allotropes of carbon.

Allotropes are different physical forms of the same solid element.

Solid carbon can existing a number of different forms. Many of these are amorphous and impure products of the burning of carbon containing fuels without adequate oxygen (e.g. charcoal from wood, coke and soot from coal, lampblack from candles). However, there are two naturally occurring pure and crystalline allotropes of carbon:

(1) *Diamond Brilliant,* colourless and hard octahedral crystals. Formed naturally of artificially at very high pressures and temperatures.

(2) *Graphite* Dull black flaky crystals formed at lower pressures. These differences in appearance result from different crystal structures.

Diamond (Fig. 8.3a) The carbon atom can form four single covalent bounds directed towards the corners of a regular tetrahedron. In a diamond crystal, each carbon atom forms a single bond with four other carbon atoms, each of which, in turn, is bonded to three others, and so on, so that all the carbon atoms in a diamond crystal are firmly bound in a regular framework. Unlike most covalently bonded substances, no molecules are formed and the bonding is uniform in all directions; in addition, there are no simple planes of weakness (compare with the NaCl structure, Fig. 8.1). Because of this crystal structure, diamond is the hardest naturally occurring substance and has a very high melting point (3500°C). Diamonds can, therefore, be used to cut other hard substances such as glass or rock (used as the 'bit' when drilling oil or water wells).

Graphite (Fig. 8.3b) The carbon atoms in graphite crystals also have a valency of four. However, each atom forms short and strong covalent bonds with three other carbon atoms and one longer and weaker bond with a fourth carbon atom. As Fig. 8.3b shows, this results in a series of flat layers of tightly bonded carbon atoms held together in stacks by weaker inter-layer bonds. We can think of each layer as being a very large two-dimensional molecule.

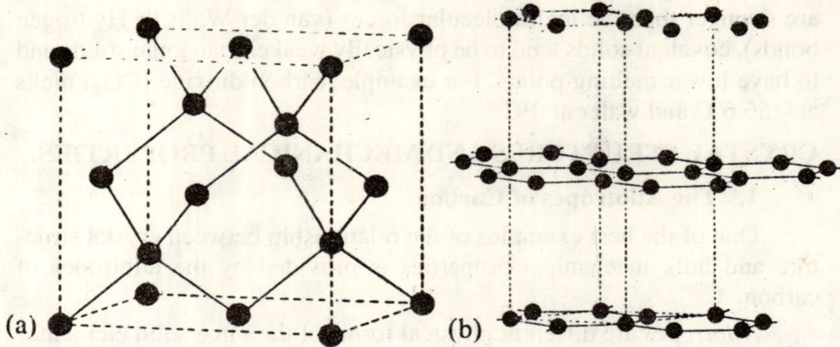

Fig. 8.3. The crystal structures of the allotropes of carbon. (a) diamond and (b) graphite.

Graphite is a soft substance due to its distinct planes of weakness which allow the layers to move over one another. Because of this property, graphite is used in pencil leads and as a lubricant, especially at high temperatures which cause the decomposition of petroleum oils. When we write with a pencil, we rub off layers of graphite which mark the paper. When graphite is used to lubricate machinery, friction is reduced by the sliding of layers carbon atoms.

Thus, according to the arrangement of the atoms, carbon crystals may be extremely hard, or soft and flaky Table 8.1

2. Metals

Iron and its alloys (the ferrous metals) are by far the most important metals used in engineering and construction. Indeed, iron accounts for more than 90% by weight of all metals used in the world today. Ferrous metals can be classed into three groups:

(*a*) Pure iron (forged iron, wrought iron, ingot iron):

(*b*) Iron containing 2.5-5% carbon (cast iron, pig iron);

(*c*) Iron containing 0.2-1.5% carbon (steels of various types).

Iron and steel production normally begins with the reduction of iron ores (naturally occurring iron oxides or carbonates) in the blast furnace. The ore is mixed with coke and limestone and heated to 1800°C. Iron is released by reduction reactions such as:

$$\underset{\text{Haematite ore}}{Fe_2O_3} + \underset{\text{Coke}}{3C} \rightarrow \underset{\text{Iron}}{2Fe} \ \underset{\text{Carbon monoxide}}{3CO}$$

It can be drawn off as the liquid metal (m.p. 1528°C) and poured into moulds to give cast iron objects. Many of the impurities in the ore, such as silicate minerals, which would impair the properties of the cast iron, are absorbed by the limestone to give a liquid material of low density called basic slag. As this floats on the surface of the liquid metal, it can be removed when withdrawing the iron. Basic slag is a useful phosphorus fertilizer for acid solids.

In spite of the refining action of limestone in the blast furnace, cast or pig iron still contains many impurities, including 2.5-5% C as well as silicon, manganese, phosphorus and sulphur compounds. In the manufacture of steel, molten pig iron from a blast furnace is purified by one of a number of different processes (Bessemer, Open hearth, etc.). The exact quantities of carbon and other ingredients are then added to give steel of the required quality. Wrought iron, which is no longer an important material, can be prepared by low temperature purification of pig iron.

The three classes of ferrous metal have distinctly different mechanical properties which are related to differences in the arrangement of their atoms. Before discussing these differences, it is necessary to present some useful definitions:

(1) *An Alloy* is a mixture of two or more metals or a mixture of a metal with a non-metal. In some cases the ingredients may react together to form an intermetallic compound.

(2) *The Hardness* of a solid substance is a measure of its resistance to scratching, wear or penetration. By observing the ability of one solid substance to scratch or cut the surface of other substances, it has been possible to draw up scales of hardness (e.g. Table 8.1). In materials testing, the hardness of a solid substance may be assessed more exactly by measuring the size of impression formed when a diamond point, is pressed into the surface of the solid.

Table 8.1 The Moh scale of hardness of minerals.

Hardness index	*Typical minerals*
1	talc
2	Gypsum, Rock Salt
3	Calcite
4	Fluorite
5	Apatite
6	Orthoclase Feldspar
7	Quartz
8	Topaz
9	Corundum
10	Diamond

Thus Graphite with a hardness between 1 and 2 is harder than talc but softer than rock salt.

(3) *The Strength* of a metal s a measure of is resistance to deformation when acted upon by a force. In particular, the *Tensile*

Strength of a metal is the maximum pulling load that a sample of the metal can endure without cracking or breaking.

(4) The *Ductility* and *Malleability* of a metal, which are measures of the case of drawing the metal into a wire or hammering it into a flat sheet, depend upon the strength of the metal. A metal with a high strength has low ductility and malleability (i.e. it resists deformation).

The mechanical properties of ferrous metals are compare qualitatively in Table 8.2; some of these differences can be explained using simple models of the structure of metals.

Most metals, including iron, are so reactive that they exist naturally as compounds and do not normally occur as the 'native' metal (exceptions are gold, silver and copper). Since the familiar metal objects we use have been shaped artificially, metals so do not, at first sight, appear to be

Table 8.2. mechanical properties of ferrous metals.

	Hardness	*Tensile Strength*	*Ductility*	*Malleability*
Pure Iron	Low	Low	High	High
Cast Iron	High	Low	Low	Low
Steel	High	High	Low	Low

crytalline, i.e. they do not have regular shapes with flat faces, straight edges and sharp vertices. However, metals are considered to be crystalline because it can be shown that in the solid state the atoms are arranged in regular three-dimensional patterns or lattices. For example, a bar of pure iron contains a large number of small iron crystals (of variable size, but typically smaller than 0.1 mm) each made up of close-packed a layers of iron atoms. If the bar is subjected to a hardness or tensile strength test, the layers of identical iron atoms will tend to slide over one another under the influence of the applied force, leading to permanent deformation of the bar (Figure 8.4). However, the resulting deformation will not be as great

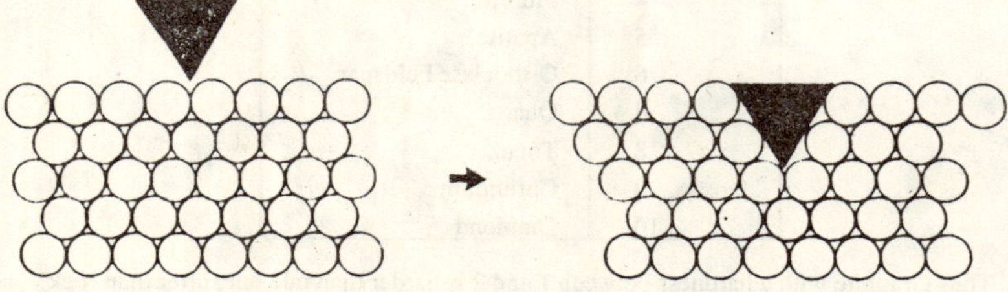

Fig. 8.4. Schematic diagram of a metal hardness test ion which a diamond point, under a standard load, causes a permanent deformation of the metal surface due to the sliding of layers of identical metal atoms.

as it would be if the bar were a single crystal, because movement will be resisted by those crystals whose layers are not aligned along the direction of sliding.

Because of its tendency of deform easily under mechanical stress, unaltered pure iron is not a useful metal in engineering where strength and hardness are important. To make the metal more useful, it is necessary to insert some 'grit' or irregularity into the structure to reduce the sliding of the layers of iron atoms.

Steels (alloys of Fe and C) are made by the addition of exact small quantities of carbon (normally 0.2-1.5%) to purified molten iron. In the liquid state and in the solid at high temperatures, the added carbon is in true solution and is distributed uniformly throughout the metal. Below 800°C the carbon comes out of solution and becomes concentrated in layers in iron carbide (Fe_3C) which are laid down between crystals of pure iron. By interleaving between layers of iron atoms, iron carbide, which is itself a very hard substance, provides the required 'grit' to increase the 'friction' between iron atoms and reduce slippage. Consequently, steel is harder and stronger than pure iron and more useful as a construction material.

In contrast, there is too much 'grit' in cast iron (up to 5%) where the presence of large fragments of graphite or iron carbide breaks up the regularity of the metal structure. Although cast iron is usually hard, it tends to be brittle, failing at the irregularities in the structure when loaded or under tension. As a result, cast iron is much less useful than steel.

In much the same way, the incorporation of a small proportion (up to 12%) of (larger) tin atoms into a sample of copper can change the metal from a soft, ductile and malleable metal used only for decorative purposes or in plumbing, to bronze, a very hard alloy, formerly widely used in making tools and weapons. The discovery of bronze, probably by the accidental smelting together of copper and tin ores, was one of Man's earliest technological advances.

Silicate Minerals in Rocks and Soils

Since the elements oxygen and silicon are by far the most common elements in the earth's crust (47% and 28% respectively, by weight), it is not surprising to find that the majority of the minerals occurring in rocks and soils are compounds of Si and O–the silicates.

Like carbon, Si is in group 4 of the periodic table; each of its atoms has four electrons in the outer shell and requires four more electrons to complete an inert gas structure (Ar). On the other hand, oxygen atoms (group 60 have six electrons in their outer shells and require a further two electrons to give the electronic structure of Ne.

The basic unit of all silicates is the Si–O tetrahedron in which a silicon atom is bound by single covalent bonds to four oxygen atoms arranged at the corner of a letrahedron–

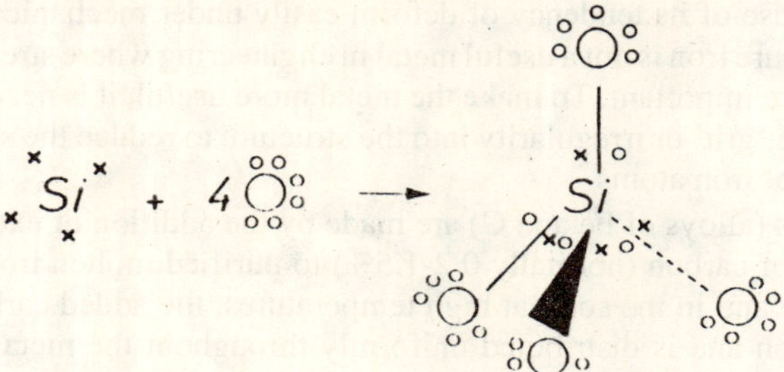

In this way, each Si atom obtains a share in a stable octet of electrons but each oxygen is still one electron short of the neon electronic structure. This deficit in electrons may be made up in a number of ways, leading to a variety of naturally occurring silicate compounds of varying physical and chemical properties.

(a) The Orthosilicates (Independent Tetrahedra)

In the orthosilicate group of minerals, the four electrons required to satisfy each Si–O tetrahedron are supplied by metal atoms, for example, in magnesium olivine, Mg_2SiO_4–

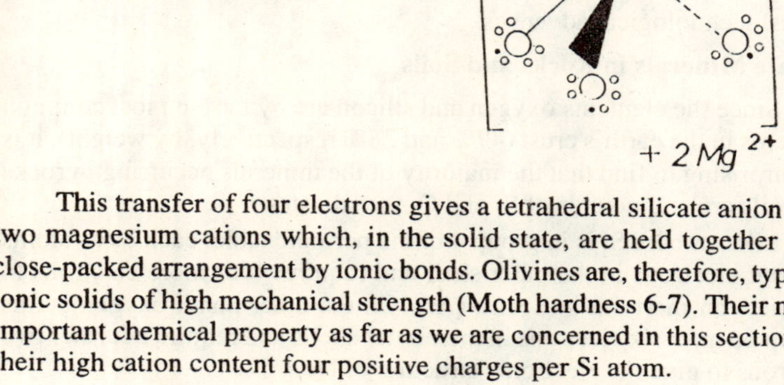

This transfer of four electrons gives a tetrahedral silicate anion and two magnesium cations which, in the solid state, are held together in a close-packed arrangement by ionic bonds. Olivines are, therefore, typical ionic solids of high mechanical strength (Moth hardness 6-7). Their most important chemical property as far as we are concerned in this section, is their high cation content four positive charges per Si atom.

(b) The Single-Chain Inosilicates or Pyroxenes (single chains of Tetrahedral)

In this group of minerals, each silicon atom shares a pair of oxygen atoms with two other silicon atoms, leading to long chains of silicate Tetrahedral bound together by Si–O–Si single covalent bonds (i.e. O atoms are shared between Tetrahedra). By this condensation of Tetrahedra, half of the oxygen atoms achieve a stable octet of electrons and only two electrons per tetrahedron are required from metal atoms. Thus, the magnesium pyroxene Enstatite, $MgSiO_3$, is made up of long chains of negatively charged Tetrahedral held together by positive magnesium ions (Fig 8.5).

Because of the very large size of the silicate 'anions' in pyroxenes the crystal structure is less regular than in orthosilicate, with the result that pyroxene minerals are not as strong and hard (Moth hardness 5-6). Pyroxenes have a cation content equal to two positive charges per Si atom.

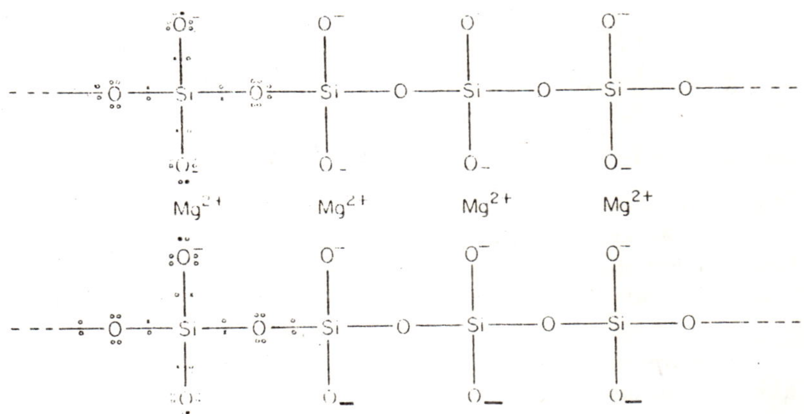

Fig. 8.5. The crystal structure of enstatite, $MgSiO_3$, showing the single chains of linked silicate tetrahedra (drawn in two dimensions) held together by magnesium ions.

(c) The double-Chain Inosilicates or Amphiboles, and the Phyllo-silicates or Sheet Silicates.

These minerals result from further condensation of silicate Tetrahedra. Amphiboles, which are made up of pairs of parallel pyroxene like chains connected by Si—O—Si bridges, resemble pyroxenes in their physical properties (Moh hardness 5–6). They normally have complex chemical compositions, e.g. Riebeckite $Na_2Fe_5\,(Si_4O_{11})_2\,(OH)_2$, but their cation content is lower than pyroxenes (equivalent to 1.75 positive charges per Si atom).

In the sheet silicates (Figure 4.6), three oxygen atoms in each tetrahedron are shared with other Tetrahedra, resulting in continuous silicate sheets held together by cations. This structure should give a cation content equivalent to one positive charger per Si atom (i.e. one electron required for the forth oxygen atom which is not shared) but the values for most sheet silicates (e.g. micas) are much higher than this due to isomorphous replacement of some Si atoms by Al atoms.

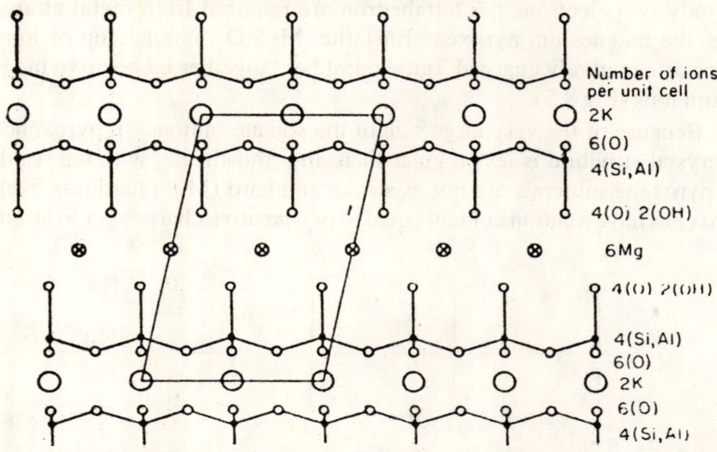

Fig. 8.6. A cross-sectional representation of an idealized phyllosilicate crystal structure, in which the sheets of silicate Tetrahedral (indicated in two dimensions by silicate *chains*) are held together by alternating layers of K^+ and Mg^{2+} ions. The *unit cell*, indicated by the enclosed area, is defined as the smallest part of a crystal structure which repeats regularly throughout the structure.

Isomorphous Replacement is the replacement of an atom (or ion) in a crystal structure by an atom of another element (or another type of ion)

Table 8.3. The radii of some metal ions (values in Angstrom units, where 1Å = 10^{-10}m) in relation to their position in the Periodic Table.

Li^+	Be^{2+}				
0.60	0.31				
Na^+	Mg^{2+}			Al^{3+}	$*Si^{4+}$
0.95	0.65			0.50	0.41
K^+	Ca^{2+}	Fe^{2+}, Fe^{3+}	Cu^+		
1.33	0.99	0.80 0.64	0.96		
Rb^+	Sr^{2+}				Sn^{4+}
1.48	1.33				0.71
	Ba^{2+}				
	1.35				

*Si is not strictly a metal (see section 1.6); the value given is for Si in silicate crystals.

without changing the crystal structure. This replacement can occur only if the atoms (ions) are of similar size (e.g.)Al for Si, Mg for Al, Table 8.3).

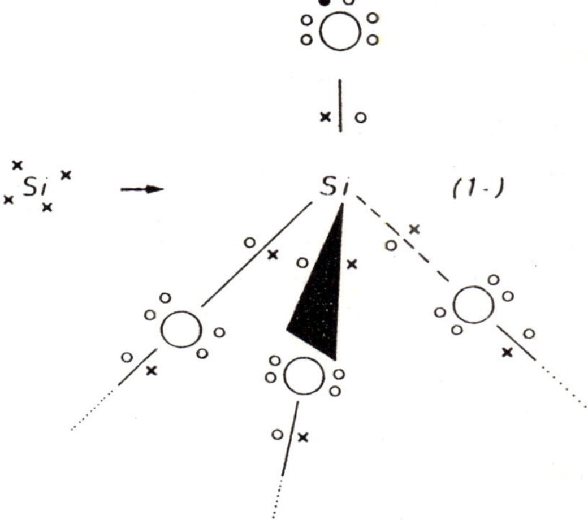

a *Normal tetrahedron in sheet silicate*

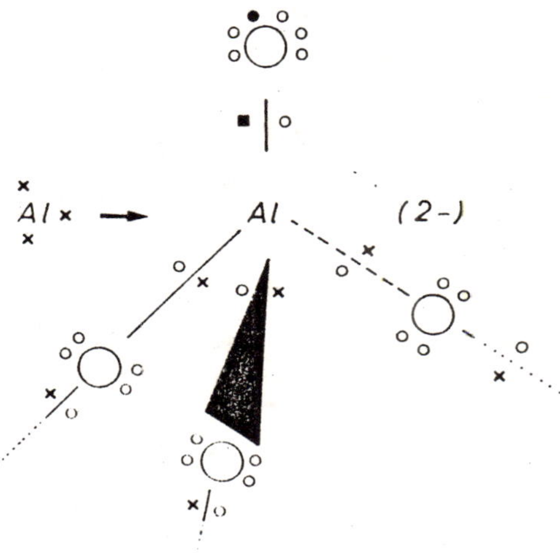

b *Tetrahedron with isomorphous replacement*

Fig. 8.7. Silicate Tetrahedra from phylosilicates (a) without isomorphous replacement, requiring one extra electron ( ) from a cation, and (b) with replacement of Si by Al. requiring two extra electrons ( ) from cations, to complete the octet of each oxygen atom.

Thus the Si atom at the centre of a silicate tetrahedron can be replaced by an Al atom whose radius is only a little greater than that of Si(Table 8.3). The space within the tetrahedron is large enough to accommodate the Al atom without great distortion of the crystal structure, but not K or Ca, for example. This substitution of Si by Al occurs to the extent of about one atom in four in micas, as indicated by their formulae:

Muscovite (white mica) $KAl_2(AlSi_3)O_{10}(OH,F)_2$

Biotite (black mica) $K(Mg, Fe)_3(AlSi_3)O_{10}(OH, F)_2$

Although Al is about the same size as Si, it is in Group 3 and, therefore, contains one less electron in its outer shell than Si. Therefore, when an Al atom replaces an Si atom in a silicate tetrahedron, it must obtain a further electron from a cation (Fig. 8.7). Thus the cation content of micas is between 1 and 2 positive charges per tetrahedron, depending upon the degree of isomorphous replacement.

Because of the slight distortion of their structures by isomorphous replacement and also because of their well-developed planes of weakness, sheet silicates tend to be flaky and physically weak (Moh hardness 2.5).

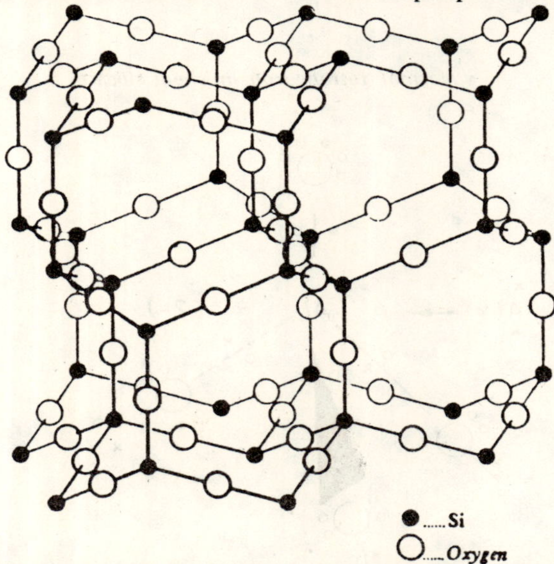

Fig. 8.8. One of the crystal structures (β-tridymite) of SiO_2. (From *Atomic Structure of Minerals,* by W.L. Bragg (1937). Cornell University Press.)

(d) The Tektosilicates or Framework Silicates

When all four of the oxygen atoms in silicate Tetrahedral are shared with adjacent Tetrahedral, then a crystal structure is formed which is covalently bonded througout and strongly resembles the structure of diamond (Fig. 8.3a and 8.4). Thus quartz (SiO_2) is mechanically strong (Moh hardness 7) and contains no cations.

However, isomorphous replacement can also occur in framework silicates leading to a reduction in strength by crystal distortion and an increase in cation content. For example, orthoclase feldspar $K(AlSi_3)O_8$, which has the same structure as quartz but with one Si atom in four replaced by Al, has a Moh hardness of 6 and a cation content equivalent to 0.25 positive charges per tetrahedron.

Table 8.4. The approximate proportions of mineral species exposed to weathering at the earth's surface.

Feldspars	30%
Quartz	28%
Micas and Clay Minerals	18%
Limestones ($CaCO_3$)	9%
Iron oxides	4%
Inosilicates	1%
Others (including orthosilicates)	10%

The Weathering of Silicate Minerals

Soils are formed by the physical, chemical an biological weathering of rocks exposed at the earth's surface, whose mineral content is given in Table 8.4. In general, the largest soil particles (sand and silt 0.002 mm diameter) are fragments of rock minerals, broken down by physical weathering (action of ice, heat, water, wind, etc.) but relatively unaffected by chemical weathering. Since resistance to physical weathering depends primarily on physical strength and hardness, we should expect from Table 8.5 that the sand and silt fractions of soils would be composed primarily of the hardest silicates –orthosilicates, feldspars and quartz.

Table 8.5. Some properties of silicate minerals.

Mineral type	*Moh handness*	*Cation content (no. of positive charges/tetrahedron)*
Olivines (orthosilicates)	6-7	4
Pyroxenes (inosilicates)	5-6	2
Amphiboles (inosilicates)	5-6	1.75
Micas (phyllosilicates)	2.5	1-2
Feldspars (tektosilicates)	6	0-25
Quartz (tektosilicates)	7	0

In contrast the clay particles (0.002 mm) in soils are predominantly new minerals synthesized during the chemical weathering of rock, sand and silt minerals. The first step in this process is the replacement of cations in silicate structures by hydronium ions (H_3O^+) causing distortion of the crystal structure and the loss of silicate Tetrahedra into solution. Thus, the cations in silicate minerals are the primary sites of chemical attack; the higher the cation content of a mineral, the more rapidly it will be weathered . Consequently, of the three hardest silicate minerals in rooks, quartz is very resistant and orthosilicates are very susceptible to chemical weathering, with feldspars intermediate.

As a result of this resistance to both physical and chemical weathering (which depends upon its crystal structure) quartz is by far the dominant mineral (up to 90%) in the sand fraction of most soils, although it accounts for a much smaller fraction of the parent rocks. Due to their susceptibility to chemical weathering, the other rock minerals tend to disappear from the sand fraction, leading to the synthesis of silicate clays. For example, an American granite which originally contained 53% feldspar and 32% quartz weathered to give a soil whose sand fraction contained 50% quartz and 28% feldspar. As weathering proceeds, the proportion of quartz will continue to increase.

CHAPTER 9

Energy and Chemistry

ENERGY

Definition of Energy

Energy is that which makes things move. It can exist in a number of forms and these forms may be changed from one to another.

Suppose that you are digging the garden. The energy to move the soil is supplied by your body. This source of energy has to be refuelled by eating food.

If you use a cultivator to dig the garden, it too needs refuelling by using petrol or connecting to the electricity supply. To generate the electricity supply, coal or oil may be burned at a power station.

If substance is being used to provide energy it is called a *fuel. Fuel* ; in one form or another, *is needed to do any useful job that involves moving.*

Thus

(a) Fuels store *chemical energy.*

(b) The *chemical energy* in coal is changed to *heat energy* when it is burnt in a power station.

(c) The *heat energy* turns water into steam which drives turbines in a dynamo.

(d) The spinning dynamo has *kinetic energy* which it converts into *electrical energy.*

(e) If *electrical energy* is used to raise a lift in a building, the lift gains *potential energy,* energy that it has by virtue of its position.

In this chapter we are concerned with chemical energy. When a chemical reaction takes place, some of this energy is usually to heat energy (occasionally light energy is produced). In fact whenever a chemical reaction occurs either *chemical energy is changed to heat energy* or *heat energy is changed to chemical energy.* The study of such changes is called **thermochemistry.**

If, when a chemical reaction occurs, *heat is evolved*, the reaction is said to be *exothermic.* If during the chemical reaction *heat is absorbed* then the reaction is said to be *endothermic.*

Energy Units

All energy is measured in *joules* (symbol used being J). A *joule* is defined *as the quantity of energy transferred when a force of 1 newton acts through a distance of 1 metre.* A joule is also equal to the energy change when 1 coulomb of electricity flows through a potential difference of 1 volt, or when 1 watt of electrical power flows for 1 second.

Heat energy used to be measure in calories, a calorie being the quantity of heat required to raise the temperature of 1 g of water by 1 °C. 4.18 joules are needed to raise to temperature of 1 g of water by 1 °C, hence 4.18 joules are equal to one calorie. The joule is such a small amount of energy that it is often convenient to use *kilojoules* (symbol: ''kJ'') (1000 J = 1 kJ).

Calories are still often used. You will for example, find the ''calorie content'' of many foods given on their packets. This is a measure of the energy content of the food when used as a fuel by you, i.e. when you eat it. You might like to work out how much energy you eat when you have, say, a dish of breakfast cereal. A source of confusion is that for food the calorie used is always the kilocalorie, i.e. 1000 calories. This is sometimes written as Calorie i.e. with a capital C, but unfortunately it is sometimes written with a small c.

The source of chemical energy

Atoms of the elements combine together to form chemical compounds. In so doing bonds are formed between the elements, and whenever a chemical bond is formed energy is released. If we were able to make chemical compounds from free atoms (that is from atoms that were not associated with any other atoms, either as a solid or liquid element or as a molecule) then we would be able to measure the decrease in the energy content of the atoms as they comes together and form a molecule and release energy. But this is usually not possible as atoms are always found with other atoms as the solid or liquid element, or combined with other atoms in molecules (except for the noble gases that do occur as free gaseous atoms). What we have to measure, therefore when a chemical reaction occurs, is *the energy between the reactants* (the chemicals that we start with), *and the products* (the chemicals formed).

We can help to explain these ideas with simple energy diagrams.

In Fig 9.1 (a) the higher horizontal line represents the total energy in some substances which react together. The lower line represents the total energy obtained from these reactants. The energy content of the products is *less* than that of the reactants so that when the reactants change to the products, *heat* is *evolved*, i.e. Fig. 9.1 (a) represents an *exothermic reaction*.

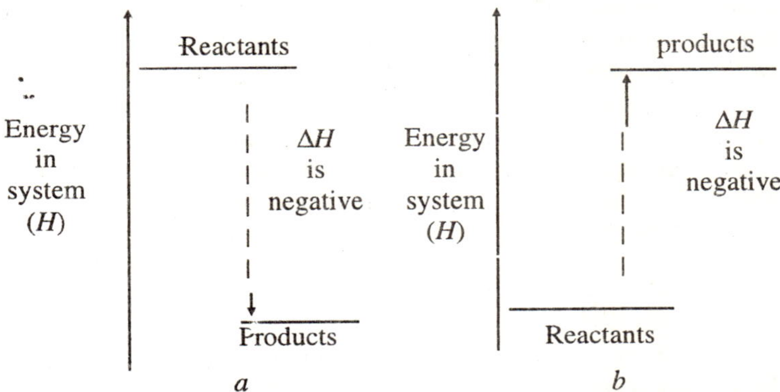

Fig 9.1 Energy content of chemical systems (a) for an exothermic reaction (b) for an endothermic reaction.

The energy content of a system (i.e. either the reactants or the products) is usually represented by the letter *H*. The *difference* between the energy content of the reactants and products is represented by ΔH (Δ is the Greek letter "delta" and means "change of"), i.e. the heat evolved in the reaction represented in Fig. 9.1 (a) is ΔH.

$$H_{\text{products}} - H_{\text{reactants}} = \Delta H$$

As H_{products} is less than $H_{\text{reactants}}$, the difference ΔH *will be negative.*

In Fig 9.1 (b) the lower horizontal line represents the total energy in the products. Here the energy content of the products is *more* than that of the reactants so that when the reactants change to the reactants change to the products, *heat is absorbed*, i.e. Fig 9.1 (b) represents an *endothermic reaction.* As before ΔH is obtained by subtracting the energy content of the reactants from the energy content of the products:

$$H_{\text{products}} - H_{\text{reactants}} = \Delta H$$

Here ΔH will be positive.

For and exothermic reaction ΔH is always negative and for an endothermic reaction ΔH is always positive.

Other names for the energy content of a system, *H*, are *heat content or enthalpy.* By convention values of ΔH given in this and other textbooks always refer to the quantity of heat evolved not by atoms and molecules but by moles of the substances involved and to emphasise this fact units are expressed as "per mole" (written mol^{-1})

ENERGY CHANGE

Energy changes on melting and boiling

When a liquid is changed into its vapour or when a solid is melted heat changes occur. Both of these changes involve the particles composing the liquid or solid gaining more freedom, i.e. bonds between the particles are broken and therefore the changes require energy.

The energy required to melt a solid is known as the *latent heat of fusion* and it is *the heat required to convert 1 g of a solid into liquid at the same temperature.* A definition involving moles is better for the chemist. The *molar heat of fusion* of a solid is *the quantity of heat required to convert 1 mole of a solid into liquid at the same temperature.*

The molar heat of fusion of ice is 6.0 kJ mol^{-1}, i.e.,

$$H_2O\,(s) \rightarrow H_2O\,(1)\ \Delta H = +6.0\ \text{kJ mol}^{-1}$$

Similarly the energy required to vaporize a liquid is known as the *latent heat of vaporization* and it is *the heat required to convert 1 g of a liquid into vapour at the same temperature.* Again, chemists prefer moles to grams, hence *the molar heat of vaporization* of a liquid is *the quantity of heat required to convert 1 mole of a liquid into vapour at the same temperature.*

The molar heat of varporization of water is 44.0 kJ mol^{-1}. This quantity of heat converts 18 g of water (i.e. 1 mole) at 100 °C into steam, i.e.

$$H_2O\,(1) \rightarrow H_2O\,(g) \quad \Delta H = +44.0\ \text{kJ mol}^{-1}$$

For fusion and vaporization, ΔH is positive. As is common experience you have to put heat into ice to melt it, or into water to boil it. These values say, therefore, that 1 mole of steam contains 44.0 kJ of energy more than 1 mole of water, and 1 mole of water contains 6.0 kJ of energy more than 1 mole of ice. The value for fusion is much less than for vaporization, because when a solid melts much of the attraction between the atoms or molecules remains, but when a liquid vaporizes all the attractions must be overcome as in the gas the molecules are widely separated.

If the processes are reversed, e.g., if we consider the condensation of steam to water, an equivalent amount of heat is liberated i.e.,

$$H_2O\,(g) \rightarrow H_2O\,(1) \quad \Delta H = -44.0\ \text{kJ mol}^{-1}$$

Latent heats of vaporization are a rough measure of the attractions existing between molecules in a liquid. For example, for the halogens the molar heats of vaporizaton are :

fluorine = + 6.7 kJ mol^{-1}

chlorine = + 17.5 kJ mol^{-1}

bromine = + 30.9 kJ mol^{-1}

iodine = + 43.5 kJ mol^{-1}

These values show that the attractive forces between the molecules must increase as the mass of the molecules increases, because more energy is required to separate them. This statement is generally true but there are exceptions.

Energy changes on burning

The most common type of chemical reaction in which heat is evolved is burning. The heat change on burning is called the *heat of combustion and it is the quantity of heat evolved when 1 mole of a substance is completely burned in oxygen.* As heat is evolved, the compound or compounds formed have a total energy that is less than the starting materials. This means that the ΔH values for combustion are always negative.

Example are:

$$Mg(s) + \frac{1}{2}O_2(g) \rightarrow MgO(s) \qquad \Delta H = -601 \text{ kJ mol}^{-1}$$

$$C(s) + O_2(g) \rightarrow CO_2(g) \qquad \Delta H = -393 \text{ kJ mol}^{-1}$$

$$CO(g) + \frac{1}{2}O_2(g) \rightarrow CO_2(g) \qquad \Delta H = -283 \text{ kJ mol}^{-1}$$

$$CH_4(g) + 2O_2(g) \rightarrow CO_2(g) + 2H_2O(1) \quad \Delta H = -890 \text{ kJ mol}^{-1}$$

To agree with the definition of heat of combustion, all of the above equations are written with *1 mole* of the *substance concerned* reacting with the appropriate amount of oxygen. The first equation, for example tells us that 1 mole of magnesium combines with $\frac{1}{2}$ a mole of oxygen gas to form one mole of magnesium oxide, and 601 kJ of energy are released for every mole (24g) of magnesium burnt.

Energy changes on dissolving

Energy is usually evolved or absorbed when a substance dissolves in water. This heat change is called the *heat of solution* and it is *the quantity of heat evolved or absorbed when 1 mole of a substance is dissolved in sufficient water so that no further heat change occurs on dilution.* There are several reasons why heat changes occur when a substance dissolves. If, for example, an ionic compound is dissolved in water the ions must be separated. To do this requires energy and dissolving is an *endothermic* process. On the other hand many ions, particularly positive ones, when they dissolve in water become hydrated, i.e., water molecules attach themselves to the ions. Hydration liberates energy and if on dissolving hydration is important the reaction in an *exotherimic* process.

Energy changes when acids and alkalis react together

The energy change that occurs when an acid is neutralized by an alkali is known as the *heat evolved when I mole of hydrogen ions (i.e., 1 g) reacts with 1 mole of hydroxide ions (i.e., 17 g), to form 1 mole of water (i.e., 18g)*:

$$H^+ (aq) + OH^- (aq) \rightarrow H_2O (1).$$

Other heats of reaction

The heat changes that we have just been considering, those that occur on combustion, on solution, and on neutralization are all heats of chemical reactions.

We can define a *heat of reactions as the quantity of heat evolved or absrobed when the number of moles represented by the normal chemical equation react together.*

THE LAW OF CONSTANT HEAT SUMMATION

This law states that *the total change in heat content during a chemical change is independent of the intermediate stages, and depends only on the chemical nature and physical state of the starting materials and products.* This law, which was discovered in the early part of the 19th century by G.H. Hess, is often called Hess's law. *The law of conservation of energy,* says that *energy can neither be created nor destroyed; but can only be converted from one form into another. Thus* Hess's law is another statement of this law as it applies to chemical change.

To illustrate the law of constant heat summation consider the formation of carbon dioxide from carbon. This can be done directly, or by forming carbon monoxide first and then burning it:

Either $$C (s) + O_2 (g) \rightarrow CO_2 (g) \qquad \Delta H = -393 \text{ kJ mol}^{-1}$$

or $$C (s) + \frac{1}{2} O_2 (g) \rightarrow CO (g) \qquad \Delta H = -110 \text{ kJ mol}^{-1}$$

followed by $$CO (g) + \frac{1}{2} O_2 (g) \rightarrow CO_2 (g) \qquad \Delta H = -283 \text{ kJ mol}^{-1}$$

The heats of reaction for the two-stage process equal the heat of reaction of the one-stage process. With this law it is obviously possible to calculate one of the heats or reaction knowing only the other two.

HEAT OF FORMATION

The heat of formation of a compound is the quantity of heat evolved when one mole of a compound is formed from its elements.

If heat is evolved when a compound is formed, e.g.,

$$Ca\ (s) + \frac{1}{2} O_2\ (g) \rightarrow CaO\ (s) \qquad \Delta H = -635 \text{ kJ mol}^{-1}$$

the compound is said to be an *exothermic compound.*

If heat is absorbed when a compound is formed, e.g.,

$$2C\ (s) + H_2\ (g) \rightarrow C_2 H_2\ (g) \qquad \Delta H = -229 \text{ kJ mol}^{-1}$$

the compound is said to be an *endothermic compound.*

Many heats of formation cannot be experimentally determined and they must be calculated using the law of constant heat summation.

In contrast the clay particles (0.002 mm) in soils are predominantly new minerals synthesized during the chemical weathering of rock, sand and silt minerals. The first step in this process is the replacement of cations in silicate structures by hydronium ions (H_3O^+) causing distortion of the crystal structure and the loss of silicate Tetrahedra into solution. Thus, the cations in silicate minerals are the primary sites of chemical attack; the higher the cation content of a mineral, the more rapidly it will be weathered . Consequently, of the three hardest silicate minerals in rooks, quartz is very resistant and orthosilicates are very susceptible to chemical weathering, with feldspars intermediate.

As a result of this resistance to both physical and chemical weathering (which depends upon its crystal structure) quartz is by far the dominant mineral (up to 90%) in the sand fraction of most soils, although it accounts for a much smaller fraction of the parent rocks. Due to their susceptibility to chemical weathering, the other rock minerals tend to disappear from the sand fraction, leading to the synthesis of silicate clays. For example, an American granite which originally contained 53% feldspar and 32% quartz weathered to give a soil whose sand fraction contained 50% quartz and 28% feldspar. As weathering proceeds, the proportion of quartz will continue to increase.

CHAPTER 10

Chemical Reactions

A chemical reaction is process which transforms one or more substances into different chemical substances. A rather dramatic example, often demonstrated in the laboratory, is the reaction of sodium metal with water; this reaction, which transforms sodium metal and water into sodium ions, hydroxide ions and hydrogen gas, also produces a great deal of heat, causing the unreacted metal to melt and rush around the surface of the water until the reaction is complete. Chemical reactions should not be confused with purely *physical processes*, like changes of state (e.g. the melting of sodium), in which the chemical nature of the substance remains unaltered.

The study of chemical reactions is normally the most important single aspect of an elementary course in chemistry. However, in agriculture and ecology, we are generally more interested in the properties and uses of inorganic chemicals and, therefore, only one chapter of Part 1 of this text is devoted specifically to reactions. In Part 2, we shall consider organic reactions in much more detail since they are the foundation of biochemistry.

CHEMICAL REACTIONS AND EQUATIONS

The chemical reaction between sodium and water can be described concisely by the chemical equation :

$$2\,Na + 2H_2O \rightarrow 2\,NaOH + H_2\uparrow$$

which is the conventional shorthand expression for :

'2 moles of sodium metal react with 2 moles of water to give 2 mole of sodium hydroxide and 1 mole of hydrogen.'

Note : (a) Since atoms are neither created nor destroyed in chemical reactions, the number of atoms of each element must be the same on each side of the equation, i.e. the equation must balance.

(b) It is conventional to avoid fractions of moles in equations. Thus the equation above is preferable to the form :

$$Na + H_2O \rightarrow NaOH + \tfrac{1}{2} H_2 \uparrow$$

(c) For reactions occurring in aqueous solution, an upward-directed arrow (e.g. $H_2 \uparrow$) indicates that a product (here H_2) is lost from solution as a gas ; similarly, a downward-directed arrow indicates the loss of a product from solution as an insoluble solid precipitate.

Since chemical reactions result from the collision of the reacting particles (atoms, ions or molecules), the faster the particles are moving, the more often they will collide with one another and react. Consequently, solid state reactions are very slow, and of little importance, because the particles are moving slowly. Most reactions of importance in biology and agricultural science occur in the liquid state and particularly in aqueous solution (cytoplasm, blood, soil solution) where the particles are moving much more rapidly. Many industrial reactions, like the production of fertilizer ammonia, are carried out in the gaseous state.

We can separate chemical reactions into a number of different classes; although it should be pointed out that a given reaction may fall into more than one class. For example :

(a) *Decomposition Reactions* involving the breakdown of a single compound into a number of simpler substances, e.g. the thermal decomposition of limestone, in lime kilns, to give quicklime (calcium oxide) :

$$CaCO_3 \rightarrow CaO + CO_2 \uparrow$$

(b) *Ionic Reactions* occurring between ions in aqueous solution (without breakdown of the ions), e.g. the exchange of anions between aluminium sulphate and calcium hydroxide (both soluble) giving two insoluble products :

$$Al_2(SO_4)_3 + 3\,Ca(OH)_2 \rightarrow 2\,Al(OH)_3 \downarrow + 3\,CaSO_4 \downarrow$$

(c) *Hydrolysis Reactions* involving the splitting of water molecules. For example, the hydrolysis of polysaccharides, lipids and proteins involve the splitting of H_2O into OH and H, which are attached to the products.

(d) *Polymerization and Condensation Reactions* linking together many similar molecules to give large macromolecules. For example the synthesis of polyethylene and polypropylene, poly saccharides and proteins.

(e) *Acid/Base Reactions.*

(f) *Redox Reactions* involving the oxidation or reduction of a substance. We shall consider redox reactions in some detail in the next two sections because of their great importance in environmental chemistry as well as their economic importance.

REDOX REACTIONS

An *oxidation* reaction occurs whenever electrons are removed from an atom or molecule, for example, in the ionic reaction :

$$2Fe^{3+} + Sn^{2+} \rightarrow 2\,Fe^{2+} + Sn^{4+}$$

Sn^{2+} ions are oxidized to Sn^{4+} ions by Fe^{3+} (ferric) ions which are the *oxidizing agents* in this reaction. The oxidation is stressed if we write the equation as a half reaction :

$$Sn^{2+} \rightarrow Sn^{4+} + 2 \text{ electrons.}$$

A *reduction reaction* occurs whenever electrons are *added* to an atom *oxidizing agents* in the reaction. The oxidation is stressed if we write the equation as a half reactions :

$$2\,Fe^{3+} + Sn^{2+} \rightarrow 2\,Fe^{2+} + Sn^{4+}$$

Fe^{3+} (ferric) ions are reduced to Fe^{2+} (ferrous) ions by Sn^{2+} ions which are the reducing agents in this reaction. Or more simply,

$$2Fe^{3+} + 2 \text{ electrons} \rightarrow 2F^{2+},$$

where the electrons originate from the oxidation of Sn^{2+} ions. In more general terms, any redox reaction can be written as follows :

$$A_{oxidized} + B_{reduced} \rightarrow A_{reduced} + B_{oxidized,}$$

where $A_{oxidized}$, in changing to Areduced, is acting as an oxidizing agent, and $B_{reduced}$, in changing to $B_{oxidized}$, is acting as a reducing agent.

This fundamental explanation of redox reactions in terms of electron exchange can be extended to demonstrate that :

(i) reactions in which oxygen is added to, or hydrogen removed from, a substance are oxidation reactions; and

(ii) reactions in which oxygen is removed from, or hydrogen added to a substance are reduction reactions.

As a simple example, the transformation of acetaldehyde (ethanal) to acetic acid (ethanoic acid) by potassium permanganate,

```
    H    O                  H    O
    |    ||                 |    ||
H — C —  C — H  →      H — C —  C — O — H
    |                       |
    H                       H
 ethanal               ethanoic acid
```

is an oxidation reaction. The added oxygen atom is highly electro-negative and, therefore, tends to electrons away from the rest of the molecule,

$$
\begin{array}{ccccc}
 & H & O & & \\
 & | & \| & & \\
H- & C- & C & \rightarrow O- & H, \\
 & | & & & \\
 & H & & &
\end{array}
$$

giving a partial *removal* of electrons from the remainder of the molecule. Thus the molecule has been oxidized.

Important Redox Reactions

(a) Redox reactions involving iron ions are very important in soil chemistry. Under aerobic conditions, the oxidation reaction,

$$Fe^{2+} \rightarrow Fe^{3+} + le,$$

is favoured and since the ferric iron is red, well-aerated soils tend to be red or reddish-brown, depending upon the organic matter content of the soil. However, under anaerobic conditions in poorly drained soils, soil microbial respiration uses ferric ions (instead of O_2) as terminal electron acceptors :

$$Fe^{3+} + le \rightarrow Fe^{2+}$$

giving bluish-grey ferrous ions. Consequently, waterlogged soils tend to be grey whereas intermittently waterlogged soils are mottled red and grey.

(b) the reduction of iron oxide in the blast furnace :

$$Fe_3\,O_4 + 4C \rightarrow 3Fe + 4CO$$

(c) Corrosion, Combustion and the Synthesis of Ammonia which are each considered in more detail in subsequent sections.

THE CORROSION OF METALS

The deterioration of metals by corrosion (rusting, etc.) is a major problem in industry and agriculture, costing many hundreds of millions of pounds each year in the UK alone. In drier climates and in areas of less severe air pollution, corrosion reactions are not so widespread but some problems are common throughout the world, for example :

(a) Rapid corrosion of the exhaust systems of motor vehicles, accelerated by the high temperatures of exhaust gases ;

(b) Rusting of unprotected metal surface exposed to moist air (for example, the corrosion of the mouldboards of ploughs can give rise to increased friction between plough and soil) ;

(c) The deterioration of metals due to reactions with acids (aerial pollutants such as SO_2, spillage of battery acid, etc.).

Most metals are highly reactive and occur in nature as oxide or sulphide ores. These ores must be reduced to give the pure metal for industrial use. For example, zinc blende (ZnS), a major zinc ore, is first roasted with oxygen to give the oxide which is then reduced by heating with coke :

$$2ZnS + 3O_2 \rightarrow 2ZnO + 2SO_2$$

$$ZnO + C \rightarrow Zn + CO$$

However, due to their reactivity, these reduced metals tend to revert to oxides and other compounds of the metal, resulting in a deterioration of the useful mechanical properties of the metal. The reactions involved in corrosion are, therefore, oxidation reactions, reversing the reduction reactions involved in their preparation. We shall examine three common types of corrosion reaction.

Formation of Oxide Films

All metals react with oxygen in the atmosphere to give a thin film of metal oxide covering the outer surface of the metal, e.g.,

$$2Cu + O_2 \rightarrow 2CuO.$$

In the case of the most useful metals for construction (e.g. Al, Ni, Zn, Fe, Cu), this oxide film is very valuable because it protects the bulk of the metal from further oxidation. Oxide films are easily demonstrated by rubbing the metal surface with sandpaper to expose bright unoxidized metal.

In other, less useful metals (e.g. K, Na, Mg, Ca), the metal oxide film tends to crack and peel off, exposing fresh metal to the atmosphere. Corrosion can, therefore, penetrate deeply into the metal and damage its mechanical properties. As we shall see below, the oxide film in iron and its alloys does not protect the metal from electrochemical corrosion, and it may be disrupted by prolonged heating. However, both of these problems can be overcome by alloying steels with Cr (13-20%), giving stainless steels whose chromium oxide film gives more complete protection from corrosion.

Electrochemical Corrosion

If pieces of two different metals are placed in contact in aqueous solution, electrons tend to flow from one metal to the other. We can predict which metal will release electrons and which will accept electrons using the *Electrochemical Series* :

Na Mg Al Ti Zn Fe Ni Sn Pb Cu Ag Pt Au,

which has been drawn up as a result of many experimental observation.

In general, electrons flow from a metal earlier in the series to a metal later in the series ; in other words, the earlier metal will be oxidized. For example, if Fe and Cu are the two metals, Fe, being earlier in the series, will supply electrons to Cu by the reaction :

$$Fe \rightarrow Fe^{3+} + 3e$$

and the electrons flowing into the copper metal combine with hydronium ions in the bathing medium to give hydrogen gas :

$$2H_3O^+ + 2e \rightarrow 2H_2O + H_2 \uparrow$$

If, instead, we have a single piece of iron containing copper as an impurity (or attached to another piece of iron by copper rivets), then, if the metal remains moist, electrons will flow from iron to copper as before. The ferric ions produced will combine with hydroxide ions in the bathing medium to give insoluble ferric hydroxide :

$$Fe^{3+} + 3OH^- \rightarrow Fe(OH)_3 \downarrow$$

This ferric hydroxide precipitate reacts further with oxygen and water to give the complex mixture of hydrated iron oxides (e.g. $Fe_2O_3 . H_2O$) known as *rust*, which flakes off the metal surface, exposing unreacted metal. Since the copper metal is not consumed in this process, the rusting of iron will continue, as long as it is exposed to water and oxygen, until it is entirely degraded to rust.

Rusting is the most familiar form of electrochemical corrosion because of the widespread use of ferrous metals and also because the iron and steel used in construction are rarely free of metallic contamination. However, electrochemical corrosion occurs whenever two metals are in contact and care must therefore be exercised if a piece of equipment is to be constructed from more than one metal or alloy.

The Prevention of Electrochemical Corrosion

The deterioration of metals by electrochemical corrosion can be prevented in two main ways.

Protective Barriers

Since both water and oxygen are required for the chemical reactions of rusting, the simplest method of preventing the corrosion of ferrous metals is to exclude both water and air from the metal surface. Long-term corrosion control can be achieved by coating the surface of metals with water-repellant paint or plastic, but layers of grease or oil can give shorter-term protection. As noted earlier, the chromium oxide surface films on stainless steels are also extremely effective.

Cathodic Protection

This method can be illustrated by galvanized iron, which is covered by a thin surface layer of zinc. If the zinc layer is damaged, allowing water to come into contract with both Zn and Fe, electrochemical corrosion will begin. However, since Zn is before Fe in the electrochemical series, it is the Zn that will be oxidized and corrosion of the iron structure will not begin until the Zn layer has disappeared. This method gives protection for at least a few years under most environmental conditions.

CORROSION BY ACIDS

If vehicle batteries are overfilled, sulphuric acid may 'slop' out of the battery, causing corrosion of the battery support and the body-word of the vehicle by the following oxidation reaction :

$$Fe + H_2SO_4 \rightarrow FeSO_4 + H_2 \uparrow$$

i.e. $$Fe \rightarrow Fe^{2+} + 2e$$

Similar corrosion reactions occur when metals are exposed to air containing high levels of acidic pollutants (SO_2, nitrogen oxides, HF, HCl etc.)

REACTIONS AND ENERGY

Chemical reactions normally involve the absorption or release of energy, usually in the form of heat (as we saw in Chapter 9). We can, therefore, class reactions as :

(a) *Exergonic*—overall release of energy. *Exothermic* if the energy released is in the form of heat; or

(b) *Endergonic*—overall absorption of energy. *Endothermic* if in the form of heat.

In biology, the most important endergonic reaction (or group of reactions) is Photosynthesis, in which the energy absorbed for the reaction is in the form of light. Overall :

$$6CO_2 + 6H_2O + \underset{\text{energy}}{\text{Light}} \rightarrow \underset{\text{carbohydrate}}{C_6H_{12}O_6} + 6O_2.$$

The absorbed energy is stored in the carbohydrate as chemical potential energy but it can be released it can be released by Respiration, an exergonic reaction (or series of reaction),

$$C_6H_{12}O_6 + 6O_2 \rightarrow 6CO_2 + 6H_2O + \text{Energy} \begin{Bmatrix} 673 \text{ Kcal mol}^{-1} \\ \text{or} \\ 2.8 \times 10^3 \text{ kJ mol}^{-1} \\ \text{for glucose} \end{Bmatrix},$$

whenever energy is required for the growth or metabolism of the plant. When the plant is subsequently eaten by a herbivore, its chemical potential energy will be used for the growth and metabolism of the animal. Thus the energy 'fixed' in photosynthesis is passed down food chains, with some less as heat at each stage. Overall, virtually all the energy required by plants, animals, man and micro-organisms originates from photosynthesis.

Photosynthesis and respiration illustrate the important generalization that a reaction which is exergonic in one direction must be endergonic in the other. (This is a necessary consequence of the Law of Conservation of Energy, i.e. that energy cannot be created or destroyed.)

A simple exergonic reaction :

$$A + B \rightarrow C + D + \text{Energy}$$

can be illustrated by Fig. 10 which stresses the fact that the reactants (A and B) have a higher energy content than the products (C and D). A clear example is the reaction of sodium metal with water :

$$2Na + 2H_2O \rightarrow 2NaOH + H_2\uparrow + \text{Energy}\begin{Bmatrix} 112 \text{ Kcal mol}^{-1} \\ \text{or} \\ 4.7 \times 10^2 \text{ kj mol}^{-1} \end{Bmatrix},$$

where the energy released is in the form of heat (melting the metal) and mechanical work (causing the molten metal to rush around the surface of the water).

In a similar way, the endergonic reaction :

$$E + F + \text{Energy} \rightarrow G + H$$

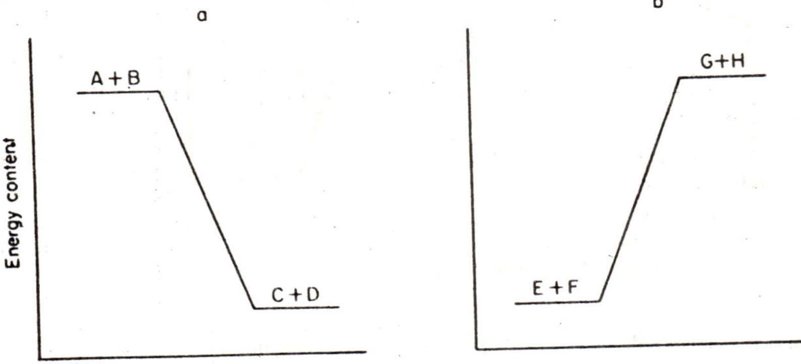

Fig. 10.1. Energy changes during (a) a simple exergonic reaction, and (b) a simple endergonic reaction.

can be illustrated by Fig. 10. 1b where the reactants (E and F) have a lower energy content than the products (G and H). Instead of 'running downhill', releasing energy, an endergonic reaction proceeds 'uphill'' and energy must be supplied. For example, the electrolysis of water,

$$2H_2O + \text{Energy} \rightarrow 2H_2\uparrow + O_2\uparrow$$

(approx. 68 kcal mol^{-1} or 2.8×10^2 kJ mol^{-1})

occurs only because electrical energy is supplied.

These diagrams illustrate the simplest situations ; however, in most reactions we must also consider the *activation energy* required, as shown in Fig. 10.2. Here the reactants must be supplied with enough energy to push them 'uphill' over the activation energy barrier before they can react. Once they have crossed the barrier, they may 'run freely downhill', returning the activation energy and also releasing more energy, normally as the Heat of Reaction. This reaction is exergonic because, although activation energy is supplied, there is an overall release of energy.

A familiar example of an exothermic reaction requiring activation energy is provided by the combustion of solid fuels. Because coal and wood have higher energy contents than carbon dioxide and water

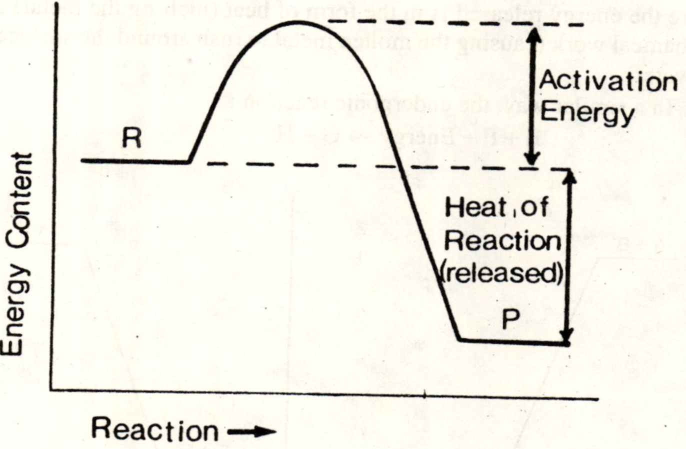

Fig. 10.2. Energy changes during an exothermic reaction requiring activation energy (R = reactants, P = products).

(the products of burning these fuels), the combustion of coal or wood releases heat which can be used for heating or cooking. However, solid fuels do not burn spontaneously, and we have to provide activation energy (heat) using burning paper and dry kindling. Once the fuel has gained enough energy to 'cross' the activation energy barrier, it will begin to burn and to release useful heat. The reaction will then continue until all the fuel has been consumed.

Note the activation energy barriers occur also in endergonic reaction (Fig. 10.3). In this case, the energy content of the reactants must be raised above that of the products before the reaction can proceed.

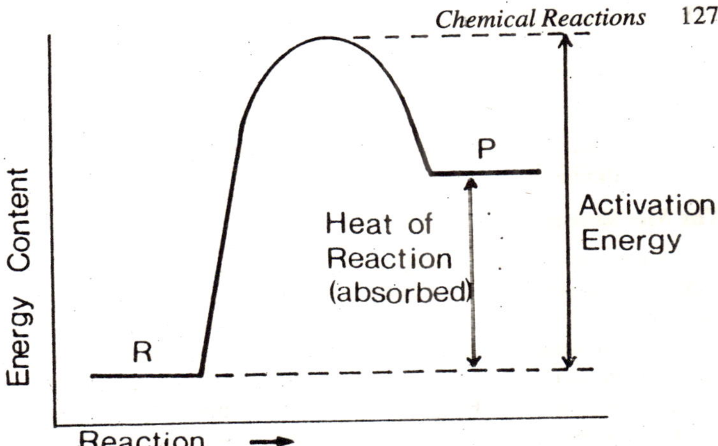

Fig. 10.3. Energy changes during an endothermic reaction requiring activation energy. Note that, in this case, only a part of the activation energy is returned as the reaction proceeds.

CATALYSTS

A catalyst is a chemical substance which, when added to a reaction mixture, changes the rate of transformation of reactants to products, without itself being used up in the process. A catalyst alters the rate of a reaction only ; it has no effect on the quantity or nature of the products obtained and it may be recovered unchanged at the end of the reaction.

Normally, catalysts are employed to speed up slow reaction. For example, in a number of gaseous reactions in the fertilizer industry, the rate of production is accelerated by finely divided metals or metal oxides.

(a) The Haber Process for the synthesis of ammonia,

$$N_2 + 3H_2 \rightarrow 2\,NH_3,$$

carried out at 300 atm pressure, 500-600 C over an *iron catalyst.*

(b) Synthesis of nitric acid (HNO_3),

First stage : $4NH_3 + 5O_2 \rightarrow 4NO + 6H_2O.$

carried out at 900° C over a *platinum catalyst.*

The subsequent reactions are :

$$2NO + O_2 \rightarrow 2NO_2$$

$$2NO_2 + H_2O \rightarrow + HNO_3 + HNO_2$$

$$3HNO_2 \rightarrow HNO_3 + 2NO + H_2O$$

(c) Synthesis of sulphuric acid (H_2SO_4),

First stage : $2SO_2 + O_2 \rightarrow 2SO_3$

carried out at 400–450° C over a *vanadium pentoxide catalyst.*

The subsequent reaction is :

$$SO_3 + H_2O \rightarrow H_2SO_4$$

Other examples include the hydrogenation of alkenes and alkynes.

The Mechanism of Catalysis

Reactions occur as a result of the collision of the reacting molecules or atoms. For example, in the Haber Process, one molecule of N_2 and three molecules of H_2 must collide together at the same instant to give the product, ammonia. Because such multiple collisions must be very infrequent when we consider the random movement of gas molecules, the rate of reaction (moles of NH_3 product per second) must be very low. The rate can be increased by raising the temperature, thus increasing the rate of movement of the molecules and the frequency of suitable collisions; This is the reason for the high temperatures used in the gas reactions mentioned above. Similarly, high pressures favour the reactions by forcing the molecules closer together. (The interactions between temperature and pressure are considered in more detail in the following section.)

The finely divided metal catalyst also increases the rate of reaction because it provides a large surface area on which molecules are adsorbed and assembled in position for the reaction. Thus the shape and great extent of the catalyst surface increase the probability that one molecule of N_2 and three molecules of H_2 will come together in a position to react.

The overall effect of a catalyst is, therefore, to lower the activation energy of the reaction, since less heat (or pressure) is required to bring the reactants into a condition to react, as shown in Fig. 10.4.

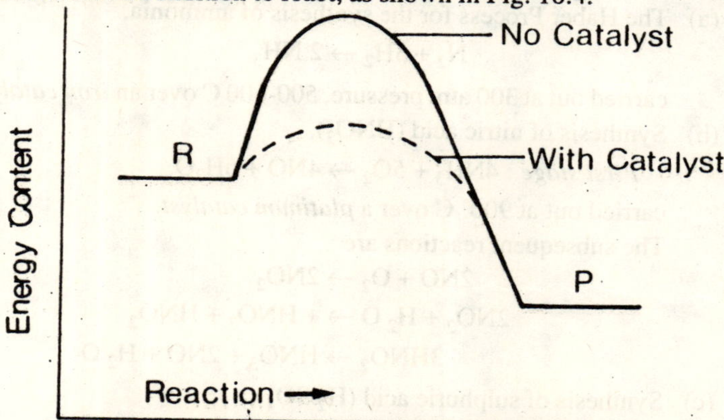

Fig. 10.4. The effect of a catalyst on an exothermic reaction requiring activation energy. The catalyst speeds up the reaction by lowering the activation energy requirement.

Catalysis is a critical factor in many industrial reactions, speeding up production and lowering energy requirements. Catalysis is even more important in biochemistry where almost all reactions involve protein catalysts called *enzymes*. Enzymes influence the rate of reactions by assembling the reactants on specific sites whose shapes are exactly defined by the tertiary structure of the protein molecule. Without enzyme catalysis, the rate of many crucial biochemical processes would be negligible.

REVERSIBLE REACTIONS

So far, we have assumed that all reactions proceed to completion and that all reactant molecules are transformed to products. However, in many cases, the reaction achieves a stable equilibrium when a certain amount of product has been formed but some reactant molecules are still present. In general terms, when the reaction

$$A + B \rightarrow C + D$$

reaches equilibrium, the reaction mixture contains A, B, C and D in constant proportions, and no more product can be obtained.

However, this stable equilibrium is not a static equilibrium in which all reactions have ceased. Instead, it is a dynamic equilibrium where the rate of the reaction forming the products

$$A + B \rightarrow C + D$$

is exactly equal to the rate of the reaction transforming these products back to the original reactants :

$$C + D \rightarrow A + B\,.$$

At equilibrium, the proportions of A, B, C and D remain constant because the rate of production of C and D is equal to the rate of consumption of C and D. This, therefore, is a reversible reaction, normally written as :

$$A + B = C + D\,.$$

For a reversible reaction, the ratio of products to reactant at equilibrium can be expressed by the *equilibrium constant* of the reaction which is derived using the *Law of Mass Action*, i.e. the rate of any chemical reaction is proportional to the (mathematical) product of the molar concentrations of the reactants. Therefore, for the reaction,

$$A + B \rightarrow C + D$$

$$\text{Rate} \propto [A]\,[B]$$

i.e. $\text{Rate}_1 = k_1\,[A]\,[B]$, where k_1 is a constant.

Similarly, for the reaction,

$$C + D \rightarrow A + B$$

$$\text{Rate} \propto [C]\,[D]$$

i.e. $\text{Rate}_2 = k_2\,[C]\,[D]$, where k_2 is a constant.

At equilibrium,

$$\text{Rate}_1 = \text{Rate}_2$$

$$k_1 [A] [B] = k_2 [C] [D]$$

and therefore, $$K = \frac{k_1}{k_2} = \frac{[C] [D]}{[A] [B]}$$

where K is the equilibrium constant for the reversible reaction.

Obviously, the higher the value of the equilibrium constant, the more product is produced by the time the reaction achieves equilibrium. Consequently, K values, which are measured under standard conditions of temperature and pressure, are very useful guides to the yield of product obtainable from a reversible reaction. For example, the Haber Process

$$N_2\ 3H_2 \rightleftharpoons 2NH_3$$

has a K value of 1.5×10^{-5} at 500° C and 1 atom pressure (if the concentrations of gases are expressed as partial pressures rather than moles 1^{-1}), indicating that very little ammonia will be formed under these conditions. In the next chapter, we shall see that equilibrium constants are very useful in expressing the strengths of acids and bases.

Although equilibrium constants are valuable in the prediction of product yield under standard condition, they cannot be used to predict how changes in conditions (temperature, pressure, etc.) will affect the equilibrium position of a reversible reaction. To do this, we must employ Le *Chatelier's Principle* which states that if the conditions of a reaction, initially at equilibrium, are changed, then the equilibrium will shift in such a direction as to tend to restore the original conditions, if such a shift is possible. We can again use the Haber Process to explain the operation of this principle.

1. Increase in Reaction Temperature

The Haber Process is an exothermic reaction :

$$N_2 + 3H_2 \rightleftharpoons 2NH_3 + \text{Heat} \begin{Bmatrix} 11\ \text{kcal mol}^{-1} \\ \text{or} \\ 4.6 \times 10^2\ \text{kJmol}^{-1} \end{Bmatrix}.$$

Consequently, if we increase the temperature of the reaction after equilibrium has been achieved, then the equilibrium position (simply the ratio of reactants to products) will change in order to remove heat and restore the original conditions. This means that the decomposition of ammonia,

$$2NH_3 + \text{Heat} \rightarrow N_2 + 3H_2,$$

will be favoured since it is endothermic, and, therefore, the yield of ammonia will decrease as the temperature rises. (Note that this effect is in opposition to the general effect or raised temperature in increasing reaction rates—see above.)

2. Increase in Pressure

The synthesis of ammonia involves a reduction in the number of moles of gas, i.e. 4 moles react to give 2, thus causing a reduction in gas pressure. Consequently, if we increase the pressure of the reaction after equilibrium has been achieved, the equilibrium position will change in order to reduce pressure and restore the original conditions. This favours the synthesis reaction :

$$N_2 + 3H_2 \rightarrow 2NH_3$$

and, therefore, the yield of ammonia will increase as the pressure rises.

3. Removal of Product

If ammonia is removed from the reaction mixture, then by Le Chatelier's Principle, the equilibrium will change so as to replace the lost product. Thus the continuous removal of ammonia will lead to greatly increased total yield.

Overall, the industrial synthesis of ammonia is carried out at high pressure in the presence of a catalyst and with constant removal of the product, ammonia. Although, in theory, the yield will be reduced by the rather high temperature employed in the reaction (500-600° C), this temperature is necessary of a reasonable reaction rate.

In this chapter, a considerable amount of space has been devoted to the Haber Process for the synthesis of ammonia because it provides an excellent illustration of the use of Le Chatelier's Principle and also because it is a major part of the manufacture of nitrogen fertilizers. However, study of the Haber Process is also important in explaining why.

Table 10.1. The energy relations of a barley crop growing in the west of Scotland.[1]

Energy inputs	*GJ per ha*[2]
Direct fuel use (machinery, etc)	1.42
Nitrogen fertilizer	4.88
Phosphorus fertilizer	0.26
Potassium fertilizer	0.31
Herbicide	0.05
Total[3]	6.92
Energy output Energy content of harvested grain[4] (Metabolizable energy)	37.8

(1) Data from J. Turnbull, quoted by M. Slessor in Chapter 1 of Food, Agriculture and the Environment. Eds. J. Lenihan and W.W. Fletcher (1975), Blackie, Glasgow.

(2) Where 1 GJ = 1×10^9 J.

(3) Excluding a number of important inputs such as grain drying, transport, processing, etc.

(4) Grain yield = 4080 kg ha^{-1} at 15% moisture content; metabolizable energy content of dry grain = 1.09×10^7 J kg^{-1}

the synthesis of nitrogen fertilizers requires such large inputs of fossil fuel to maintain the high temperatures (500—600° C) and pressures (300 atm) necessary for an economic yield. Consequently, the amount of energy (in the form of oil, coal, etc.) required to produce the nitrogen fertilizer for a hectare of cereals under mechanized agriculture, can be much larger than the energy required for all other purposes (tillage, pesticides, harvesting, etc.), and it may be equivalent to a significant fraction of the energy content of the harvested crop (derived from solar radiation) (Table 10.1).

CHAPTER 11

Acids and Bases

An Acid is a hydrogen-containing substance which dissociates when dissolved in water, to give hydronium (H_3O^+) ions, e.g.

Sulphuric acid

$$H_2SO_4 + 2H_2O \rightarrow 2H_3O^+ + SO_4^{2-}$$

Acetic acid

$$CH_3CO_2H + H_2O \rightarrow H_3O^+ + Ch_3CO_2^-.$$

As these examples indicate, the dissociation of an acid in water in volves the transfer of a proton (H^+) from the acid to a water molecule. Consequently, acids are known as *proton donor* substances.

All acid have certain properties in common, in addition to raising the hydronium ion concentration of aqueous solution. These include a sour or sharp taste, a tendency to 'burn' the skin, the capacity to dissolve metal and to neutralize bases (see below).

A *Base* is a substance, commonly containing the hydroxyl group or ion, which ionizes in water to give hydroxide (OH^-) ions, e.g.

potassium hydroxide $$KOH \rightarrow K^+ + OH^-$$

ammonia $$NH_3 + H_2O \rightarrow NH_4^+ + OH^-$$

In contrast to acids, bases are *proton acceptor* substances. This characteristic is clearly shown in the ionization of ammonia, but, as we shall see in the next section, it is a feature of all bases including hydroxides like KOH. Bases have 'brackish' tastes and feel soapy. The most soluble bases (NaOH, KOH) are called *alkalis*.

In the following sections we shall investigate some of the properties of acids and bases which make them of paramount importance in biochemistry and soil science.

ACIDITY, ALKALINITY AND pH

The acidity of a solution is the concentration of hydronium ions in the solution, whereas the alkalinity of a solution can be defined as the concentration of hydroxide ions. In a given solution, these two concentration values cannot vary independently ; for example, when $[H_3O^+]$ is high, $[OH^-]$ must be low, and vice versa. The reason for this inverse relationship is as follows : The dissociation of water is a reversible reaction :

$$H_2O + H_2O \rightleftharpoons H_3O^+ + OH^-,$$

whose equilibrium constant is given by :

$$K = \frac{[H_3O^+][OH^-]}{[H_2O]^2}$$

$[H_2O]$, the molar concentration of water in water (55.6 M) is obviously a constant and, therefore, $[H_2O]^2$ can be absorbed into the equilibrium constant to give a new constant :

$$K_w = [H_3O^+][OH^-],$$

where K_w (the Ion Product of water) has been found by experiment to be 10^{-14} (at 25° C).

In other words, the mathematical product of the hydronium ion concentration and the hydroxide ion concentration in any aqueous solution at 25° C will be equal to 10^{-14}. For example, if the hydroxide ion concentration in a given solution is 10^{-4} M (0.0001 M), then the hydronium ion concentration will be 10^{-10} M (0.000 000 000 1 M) as shown below :

$$[H_3O^+] = \frac{K_w}{[OH^-]} = \frac{10^{-14}}{10^{-4}} = 10^{-10}$$

Similarly, if the hydronium ion concentration rises from 10^{-10} M to 10^{-4}, M, the hydroxide ion concentration will fall from 10^{-4} M to 10^{-10} M.

Using this interrelationship, we can show that solutions are acidic (contain more hydronium than hydroxide ions) when the hydronium ion concentration is higher than 10^{-7} M and alkaline when the hydroxide ion concentration exceeds 10^{-7} M. A neutral solution (neither acidic nor alkaline) contains equal concentrations of H_3O^+ and OH^-, both 10^{-7} M. (we can now see why bases are called proton acceptors. When a base dissolves in an aqueous solution, it causes an increase in hydroxide ion concentration when must be accompanied by a reduction in hydronium ion concentration according to the neutralization reaction :

$$H_3O^+ + OH^- \rightarrow 2H_2O.$$

Here each OH^- ion is accepting a proton from a hydronium ion.)

The acidity and alkalinity of some important aqueous solutions are as follow :

	Acidity $[H_3O^+]$	*Alkalinity* $[OH^-]$
Gastric fluid (pig)	10^{-2} M	10^{-12} M
Saliva (human)	10^{-7}	10^{-7}
Saline soil solution	down to 10^{-5}	up to 10^{-5}

In order to avoid clumsy and troublesome acidity values like 0.000 000 1 M or 10^{-7} M, it has become conventional to convert them to pH values, which fall conveniently within a 0 to 14 scale.

The pH of a solution is defined by the equation :

$$pH = -\log_{10}[H_3O^+].$$

Therefore, for pig gastric fluid :

$$pH = -\log_{10}(40^{-2}) = -(-2)$$
$$= 2.$$

Similarly, the pH of human saliva is 7 and the pH of saline soil solution can be as high as 9.

Using the definition :

$$pH = -\log_{10}[H_3O^+]$$

and the relationship :

$$K_w = [H_3O^+][OH^-] = 10^{-14}$$

we can establish the relationships between pH, $[H_3O^+]$ and $[OH^-]$ shown in Table 11.1 in using the pH scale, the following points should be stressed :

(a) as the pH value rises, $[H_3O^+]$ falls and $[OH^-]$ rises ;

(b) the pH of a solution defines both the $[H_3O^+]$ and the $[OH^-]$ concentrations of the solution ;

(c) a difference of one unit of pH indicates a tenfold difference in $[H_3O^+]$ concentration ;

(d) It is not normally necessary to quote pH values to more than one decimal place (particularly since laboratory pH meters are scarcely accurate to 0.1 of a pH unit).

Table 11.1 The relationships between pH, H_3O^+ and OH^- in aqueous solutions.

	Acid solution							Alkaline solutions					
pH	1	2	3	4	5	6	7	8	9	10	11	12	13
$[H_3O^+]$	10^{-1}	10^{-2}	10^{-3}	10^{-4}	10^{-5}	10^{-6}	10^{-7}	10^{-8}	10^{-9}	10^{-10}	10^{-11}	10^{-12}	10^{-13}
	← increasing												
$[OH^-]$	10^{-13}	10^{-12}	10^{-11}	10^{-10}	10^{-9}	10^{-8}	10^{-7}	10^{-6}	10^{-5}	10^{-4}	10^{-3}	10^{-2}	10^{-1}
	increasing →												

The Importance of pH

Enzymes

The catalytic activity of enzymes in biochemical reactions depends upon the binding of reactant molecules to sides which are specifically shaped to accommodate these molecules only. However, by altering the ionization of amine base, carboxylic acid and other functional groups, variations in pH can cause changes in the distribution of charge on the surface of enzyme molecules which, in turn, alter the ability of enzymes to bind reactant molecules. Consequently, each enzyme has an optimum pH range for reactant binding, and the medium in which enzyme catalysis occurs is held precisely within that pH range by means of buffers. For example, the digestive enzyme in human saliva operates optimally at pH 7 whereas the digestive enzymes in the human stomach gastric fluid, pH 2) have an optimum pH values are controlled to permit enzymes to act most efficiently include :

Bovine rumen	5.5—6.5
Vertebrate blood and lymph	7.4
Cell cytoplasm	6.9

Soils

The pH of a soil influences the solubility of phosphate, trace elements (Fe, Mn, Zn, Cu, Co, Mo) and toxic ions (e.g. Al^{3+}) as well as controlling the activity of soil micro-organisms. For example, the fixation and mineralization of nitrogen by free-living bacteria cease at soil pH values below 4. Because of these effects, the pH values of agricultural soils are normally maintained within the range 5.5 to 7.0 to ensure adequate supplies of phosphate and trace elements, and active microbial populations, coupled with low solubilities of toxic ions. However, plant species can colonize soils whose pH values lie outside this range, if they have evolved adaptations to overcome the adverse factors associated with extremes of pH (e.g. trace element deficiencies at pH > 7).

pH is a crucial factor in many other aspects of agricultural science; to give one other example, the pH of freshly prepared silage must fall rapidly to 4.0, otherwise undesirable micro-organisms (e.g. *Clostridium* spp.) will invade the mixture, giving a poor-quality or toxic silage.

THE STRENGTH OF ACIDS AND BASES

We can separate acids into two classes according to their ionization when dissolved in water, i.e. :

(*a*) *Strong Acids* which ionize (dissociate) completely in water, e.g. :

$$HNO_3 + H_2O \rightarrow H_3O^+ + NO_3^-$$

(Nitric acid)

$$H_2SO_4 + 2H_2O \rightarrow 2H_3O^+ + SO_4^{2-}$$

(Sulphuric acid)

These ionization reactions go to completion and are not reversible. In a solution of nitric acid, for example, there are H_3O^+ and NO_3^- ions only and (almost) no undissociated HNO_3.

(*b*) *Weak Acids* which do not ionize completely in water, e.g. :

$$CH_3CO_2H + H_2O \rightleftharpoons H_3O^+ + CH_3CO_2^- \text{ (Acetic Acid)}$$

$$\left.\begin{array}{l} H_3PO_4 + H_2O \rightleftharpoons H_3O^+ + H_2PO_4^- \\ H_2PO_4^- + H_2O \rightleftharpoons H_3O^+ + HPO_4^{2-} \\ HPO_4^{2-} + H_2O \rightleftharpoons H_3O^+ + PO_4^{3-} \end{array}\right\} \text{(Phosphoric acid)}$$

These ionization reactions do not go to completion and are, therefore, reversible reactions. A solution of acetic acid contains H_3O^+ ions $CH_3CO_2^-$ ions and undissociated CH_3CO_2H.

Consequently, a 1 M solution of weak acid has a lower concentration of hydronium ions (i.e. a higher pH) than a 1 M solution of a strong acid. (It is important to stress at this point that the strength of an acid should not be confused with the pH of an acidic solution.)

In the same way, bases can be classed as :

(*i*) *Strong Bases* (the alkalis) which dissociated completely in water, e.g. :

$$KOH \rightarrow K^+ + OH^-$$

(*ii*) Weak Bases which do not dissociate completely, e.g. :

$$Ca(OH)_2 \rightleftharpoons Ca^{2+} + 2OH^-$$

Thus, an aqueous solution of calcium hydroxide contains Ca^{2+} ions, OH^- ions and undissociated $Ca(OH)_2$.

We can express the strength of an acid (i.e. the extent of its ionization) using the equilibrium constant of its ionization. Thus for the ionization of acetic acid :

$$CH_3CO_2H + H_2O \rightleftharpoons H_3O^+ + CH_3CO_2^-$$

$$K = \frac{[H_3O^+][CH_3CO_2^-]}{[CH_3CO_2H][H_2O]}$$

Since $[H_2O]$ is a constant, it can be absorbed into K to give a new constant,

$$K_a = \frac{[H_3\,O^+]\,[CH_3\,CO_2^-]}{[CH_3\,CO_2\,H]}$$

and since K_a (the Dissociation Constant) is a ratio of ionized to unionized acetic acid, it can be used as an index of the strength of the acid. K_a values for a selection of acids are given in Table 11.2 ; the lower the value of K_a, the weaker the acid.

The dissociation constant, K_a, of an acid indicates the extent of ionization of the acid when it alone is dissolved in water. However, Table 11.2 can also be used to predict the outcome of dissolving two (or more) acids in the same solution. In all cases, the stronger acid will ionize according to its K_a value but the ionization of the weaker acid will be suppressed. Thus, in a solution containing equal concentrations of hydrochloric and acetic acids, HCl will dissociate completely to give H_3O^+ and Cl^- ions whereas the acetic acid will exist almost entirely as the undissociated acid $CH_3\,CO_2\,H$.

Table 11.2. Dissociation constants of some common acid 25° C).

Dissociation	K_a	pK_a
$HCl + H_2\,O \rightarrow H_3\,O^+ + Cl^-$ (Hydrochloric Acid)	10^7	strong acid
$H_3\,PO_4 + H_2\,O \rightleftharpoons H_3\,O^+ + H_2\,PO_4^-$	7.5×10^{-3}	2.1
$CH_3\,CO_2\,H + H_2\,O \rightleftharpoons H_2\,O^+ + CH_3\,CO_2^-$	1.8×10^{-5}	4.8
$H_2\,CO_3 + H_2\,O \rightleftharpoons H_3\,O^+ + HCO_3^-$ (Carbonic Acid)	4.3×10^{-7}	6.4
$H_2\,PO_4^- + H_2\,O \rightleftharpoons H_3\,O^+ + HPO_4^{2-}$	6.2×10^{-8}	7.2
$C_6\,H_5\,OH + H_2\,O \rightleftharpoons H_3\,O^+ + C_6\,H_5\,O^-$ (Phenol)	1.3×10^{-10}	9.9
$HCO_3^- + H_2\,O \rightleftharpoons H_3\,O^+ + CO_3^{2-}$ (Bicarbonate)	5.6×10^{-11}	10.3
$HPO_4^{2-} + H_2\,O \rightleftharpoons H_3\,O^+ + PO_4^{3-}$	2.2×10^{-13}	12.7

Another use of the dissociation constant can be explained as follows. If we imagine that the dissociation of acetic acid molecules in aqueous solution has been suppressed by a very low pH (i.e. presence of a strong acid), then as we raise the pH, the molecules begin to dissociate to give acetate ions. At a certain pH, half of the acid molecules will have dissociated to give acetate ions, i.e.

$$[CH_3\,CO_2\,H] = [CH_3\,CO_2^-].$$

Therefore, the equation :

$$K_a = \frac{[H_3\,O^+]\,[CH_3\,CO_2^-]}{[CH_3\,CO_2\,H]}$$

simplifies to give

$$K_a = [H_3\,O^+]$$

i.e. the dissociation constant of the acid is equal to the hydronium ion concentration at which the aid is 50% ionized. Taking negative logarithms we obtain

$$-\log_{10} k_a = -\log_{10}[H_3\,O_+]$$

or

$$p\,K_a = pH$$

i.e. the pK_a of an acid is the pH at which it is 50% ionized. Some pK_a values are given in Table 11.2.

This relationship allows us to begin to predict the ionic form in which acids exist in biological fluids, soil solutions and other aqueous media. For example, since the ionization :

$$H_2\,PO_4^- + H_2\,O = H_3\,O^+ + HPO_4^{2-}$$

has a pK_a of 7.2, phosphoric acid added to vertebrate blood (pH 7.4) will ionize to give approximately 50% $H_2\,PO_4^-$ ions and 50% HPO_4^{2-} ions. HPO_4^{2-} will increasingly predominate as the pH rises above 7.4, whereas the concentration of $H_2\,PO_4^-$ ions is higher below pH 7.4. More complete calculations give the proportions of the four species ($H_3\,PO_4$, $H_2\,PO_4^-$, HPO_4^{2-}, PO_4^{3-} shown in Fig 11.1.

The strength (degree of ionization) of bases can also be expressed using dissociation constant, e.g. for NH_3 :

$$NH_3 + H_2\,O \rightleftharpoons NH_4^+ + OH^-$$

$$K = \frac{[NH_4^+]\,[OH^-]}{[NH_3]\,[H_2\,O]}$$

and, therefore,

$$K_b = \frac{[NH_4^+]\,[OH^-]}{[NH_2]}$$

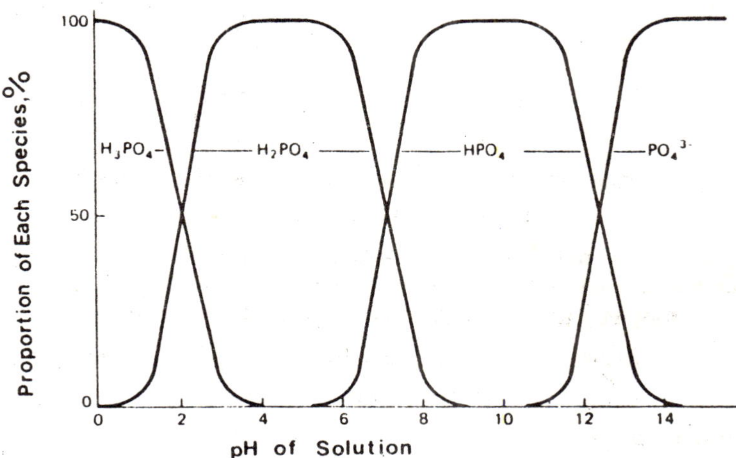

Fig. 11.1. Proportion of phosphoric acid and its ions at different pH values. Note that over the pH range of most agricultural soils (5-7), the predominant ion in solution is $H_2PO_4^-$.

Dissociation constants for a selection of amine bases are presented in Table 17.1. ; the lower the value of K_b, the weaker the base.

The pH of a soil is a measure of the hydronium ion concentration in the soil solution. Most of these hydronium ions originate from ionization reaction, e.g. I, where the 'colloid' in the equation represents a clay or humus particle in the soil. The question suggests only two ionizable hydrogens per colloid

I

$$\text{Colloid}{<}^{H}_{H} + 2H_2O \rightleftharpoons 2H_3O^+ + \text{Colloid}$$

(weak acid)

particle, for clarity, although each particle will, in practice, carry a large number.

The pH resulting from such dissociation dependents upon.

(a) a representative K_a value (difficult to assign a meaningful value), and

(b) the number of ionizable hydrogen atoms on the clay and humus.

Consequently, any treatment which reduces the number of ionizable hydrogen atoms attached to the soil colloids will tend to raise the pH of the soil. This is the reason for treating acidic soils with lime

($CaCO_3$, CaO or Ca $(OH)_2$) ; Ca^{2+} ions replace the hydrogen atoms on the colloid by the ion exchange reaction (II), and the resulting hydronium

II

$$\text{Colloid}{<}^{H}_{H} + Ca^{2+} + 2H_2O \rightleftharpoons \text{Colloid}{>}Ca + 2H_3O^+$$

ions are neutralized by hydroxide ions, also supplied by the lime :

$$2H_3O^+ + 2OH^- \rightarrow 4H_2O.$$

The overall result of these reactions will be a rise in soil pH. Ion exchange reactions are considered in more detail in Chapter 12.

EQUIVALENT WEIGHTS AND NORMALITY

— Acids and bases react together by neutralization reactions, e.g. :

$$NaOH + HCl \rightarrow H_2O + NaCl$$

or more simply

$$OH^- + H_3O^+ \rightarrow 2\,H_2O.$$

Consequently 1 litre of 1 M NaOH will neutralize 1 litre of 1 M HCl. However, in the reaction :

$$2NaOH + H_2SO_4 \rightarrow Na_2SO_4 + 2H_2O,$$

2 litres of 1 M NaOH are required to neutralize 1 litre of 1 M H_2SO_4, because H_2SO_4 molecules contain two ionizable hydrogen atoms.

These example illustrate the advantage of using *gram equivalent weights* (mass of each substance yielding 1 g of protons or 17 g of hydroxide ions) rather than gram molecular weights in acid/base reactions. In formal terms :

The gram equivalent weight *GEW* of an acid is the gram molecular weights divided by the number of ionizable hydrogen atoms in the formula.

Thus for HCl, the *GEW* is $\frac{36.5}{1} = 36.5$ g

for H_2SO_4, $GEW = \frac{98}{2} = 49$ g

for CH_3CO_2H, $GEW = \frac{60}{1} = 60$ g (only 1 H ionizable)

The gram equivalent weight of a base is the gram molecular weight divided by the number of hydroxide ions produced per 'molecule'.

Thus for NH_3, $GEW = \frac{17}{1} = 17$ g

$$\text{for Ca (OH)}_2,\ GEW = \frac{74}{2} = 37 \text{ g}$$

$$\text{for NaOH,}\quad GEW = \frac{40}{1} = 40 \text{ g}$$

Thus 1 litre of NaOH solution, containing 1 gram equivalent weights (40 g) per litre will neutralize 1 litre of $H_2\ SO_4$ solution, containing 1 gram equivalent weight (49 g) per litre. Or, more concisely, 1 litre of 1 N NaOH will neutralize 1 litre of 1 N H_2SO_4 where *1 N* (*1 Normal*) solution contains 1 gram equivalent weight/litre.

Gram equivalent weights and normality are also used for ionic solutes in ion exchange and redox reactions. In ion exchange reactions, ionic charge is of paramount importance, and, therefore, the gram equivalent weight is obtained simply by dividing the gram ionic weight by the charge on the ion, e.g.

$$\text{for } H_3\,O^+,\ GEW = \frac{19}{1} = 19 \text{ g}$$

$$Ca^{2+},\ GEW = \frac{40}{2} = 20 \text{ g}$$

$$Fe^{3+},\ GEW = \frac{56}{3} = 18.7 \text{ g}$$

$$SO_4^{2-},\ GEW = \frac{96}{2} = 48 \text{ g.}$$

However, in redox reactions, the number of electrons *transferred* is more important that the value of the charge on the ion ; consequently, for Fe^{+3}, which can accept one electron by the redox half reaction :

$$Fe^{3+} + 1e^- \rightarrow Fe^{2+},$$

$$GEW = \frac{56}{1} = 56 \text{ g.}$$

Thus for ion exchange reactions, *GEW* is a measure of the mass of ion providing unit electrical charge whereas, in redox reactions, it is a measure of the mass of ion exchanging unit quantity of electrons ; since these definitions can result in different *GEW* values for the same ion (e.g. Fe^{3+} above), it is essential to use the correct procedure consistently for each application.

BUFFERS

The pH of biological fluids must be controlled precisely to ensure efficient enzyme catalysis. For example, human blood is maintained within 0.1 of 7.4 since a variation of as little as 0.2 could prove fatal. In nature, steady pH values are usually achieved using a buffer system, based on the dissociation of a weak acid, which 'mops up' any H_3O^+ and OH^- ions produced by the metabolism.

The mechanism of buffering can be illustrated by the carbonic acid/bicarbonate ion buffer system. In solution, we have the dynamic equilibrium :

$$H_2CO_3 + H_2O \rightleftharpoons HCO_3^- + H_3O^+$$

If we disturb this equilibrium by adding H_3O^+ ions (i.e. lowering the pH), then by Le Chatelier's Principle the equilibrium position will change so as to remove the added H_3O^+ ions. Therefore, the reaction :

$$HCO_3^- + H_3O^+ \rightarrow H_2CO_3 + H_2O$$

will proceed until all the added H_3O^+ ions have been removed from solution and the original pH is re-established.

If instead, we add OH^- ions, they will react with H_3O^+ ions :

$$H_3O^+ + OH^- \rightarrow 2H_2O$$

thus raising the pH. In this case, according to Le Chatelier's Principle the ionization reaction :

$$H_2CO_3 + H_2O \rightarrow HCO_3^- + H_3O^+$$

will be favoured until the original pH has been regained. This 'mopping-up' of H_3O^+ and OH^- ions can occur only if both elements of the system are present in solution, i.e. the anion and the undissociated weak acid.

The pH of the buffered solution is related to the pKa of the appropriate weak acid as follows (using the carbonic acid/bicarbonate system as an example) :

$$K_a = \frac{[H_3O^+][HCO_3^-]}{[H_2CO_3]}$$

Therefore,

$$[H_3O^+] = K_a \frac{[H_2CO_3]}{[HCO_3^-]}$$

and

$$pH = pK_a - \log \frac{[H_2CO_3]}{[HCO_3^-]}$$

$$= pK_a + \log \frac{[HCO_3^-]}{[H_2CO_3]}.$$

For example, human blood, containing 0.025 M bicarbonate and 0.00125 M carbonic acid is buffered at :

$$pH = 6.1^* + \log\left(\frac{0.025}{0.00125}\right)$$
$$= 7.4.$$

This buffer system is principally employed in removing hydronium ions generated by tissue metabolism, with the excess carbonic acid produced by the buffer being lost from the lungs as CO_2, as a result of the enzyme-catalyzed reaction :

$$H_2CO_3 \rightarrow H_2O + CO_2\uparrow$$

Thus the bicarbonate ion concentration remains very much higher that the carbonic acid concentration, thereby ensuring a steady blood pH.

AMPHOTERIC SUBSTANCES

Certain substances, notably the amino acid and proteins can act either as acids (proton donors) or as bases (Proton acceptors) under different circumstances. Such substances, which can contribute to the pH-buffering of living tissues, are called *amphoteric*. For example, the α-amino acid glycine has a carboxylic acid group (weak acid) at one end of the molecule and an amine group (weak base) at the other :

$$NH_2 - CH_2 - CO_2H$$

When glycine is dissolved in water at near-neutral pH, both groups dissociate to give a *zwitterion* (charged at both ends) :

$$NH_3^+ - CH_2 - CO_2^-$$

As the pH of the solution is lowered, the ionization of the carboxylic acid group is suppressed and the molecule acts as a portion acceptor :

$$NH_3^+ - CH_2 - CO_2^- + H_3O^+ \rightarrow NH_3^+ - CH_2 - CO_2H + H_2O,$$

giving a positively charged ion.

IN contrast, as the pH is raised, the ionization of the amine group is suppressed and the molecule acts as a proton donor :

$$NH_3^+ - CH_2 - CO_2^- + OH^- \rightarrow NH_2 - CH_2 - CO_2^- + H_2O,$$

giving a negatively charged ion.

These changes can be detected using an electrophoresis apparatus, in which a solution of the substance under test is subjected to an electric field. At low pH values, the glycine molecules (as a positive ion) moves towards the negative pole (cathode), whereas at high pH values, the molecule (negative ion) moves towards the anode. At a definite intermediate pH value (6.1), called the *isoelectric point*, the glycine molecule does not move in either direction because the charges on each end of the molecule are the same. Since each of the 22 important α-amino acids has a different isoelectric point (due to the influence of the side-chains), electrophoresis is a useful tool in the identification of amino acids.

CHAPTER 12

Water

Water is the major constituent of both plants (over 90% in annuals) and animals (typically 60-70% in vertebrates). In living tissues, water is the medium for many biochemical reactions and excretion process ; inorganic nutrients, photosynthate, gases and hormones are all transported in aqueous solution ; evaporation of water can control the temperature of skin or leaf ; soil nutrients are available to plant roots only when dissolved in soil water. In short, water is essential for life and plays a unique role in virtually all biological processes.

In addition to this, water fulfils several other roles in agriculture. For example, most pesticides are applied to crops or livestock as aqueous solutions or suspensions, and vast quantities of water are used in irrigation and fish culture. Water also plays a central part in soil formation and erosion.

The paramount importance of water in agricultural production can be stresses by the following facts :

(a) it has been estimated that between 35 and 45 metric tonnes of water are required for the production of each kilogram of beef in the USA (this includes water for all purposes—fodder production, dipping, carcase preparation, etc., as well as the physiological requirements of the animal).

(b) Of the soil moisture absorbed by a maize crop, 98% is lost by transpiration ; the remainder is retained in the plant body and only 0.2% is chemically broken down in photosynthesis.

In this chapter we shall see that although water is a very familiar substance it has unusual properties which make it uniquely suited to fulfil its many roles.

THE STRUCTURE OF LIQUID WATER

The water molecule is planar and angled and that, due to the high electronegativity of oxygen, there is a partial separation of charge (I) leading to the formation of hydrogen bonds between molecules (II).

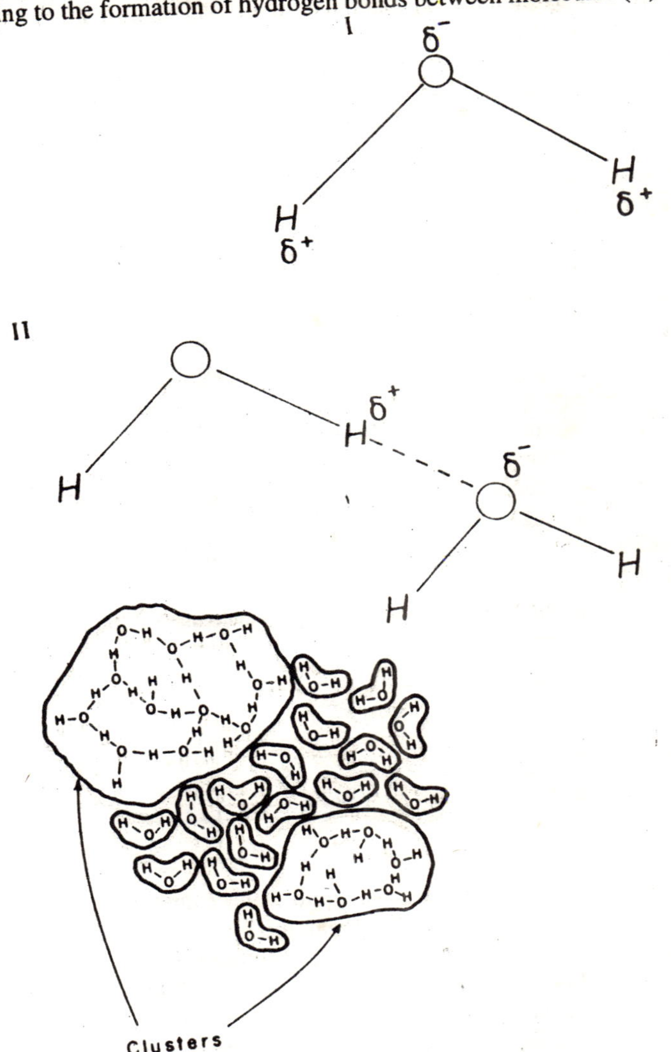

Fig. 12.1 The 'Flickering Cluster' structure of liquid water, in which 'free' water molecules are continually exchanging with hydrogen-bonded molecules in the clusters.

Note that although these hydrogen bonds are strong compared with other intermolecular forces (van der Waals bonds), they are only about 1/24th as strong as an O—H covalent bond.

If this system of hydrogen bonding between adjacent water molecules were to extend uniformly throughout liquid water, it would be solid (ice) at room temperature, with a regular crystal structure. Clearly, liquid water is not a solid at room temperature and, therefore, it is generally thought that it is made up of a large number of small clusters (or 'icebergs') of hydrogen-bonded water molecules separated by volumes containing free unbonded molecules only (Fig. 12.1). Each of these clusters, which have a rather open, cage-like structure, is continually exchanging water molecules with other clusters and with the pool of unbonded molecules. This 'flickering cluster' structure is, therefore, dynamic, in contrast to the static structure of crystalline solids.

THE THERMAL PROPERTIES OF WATER

When a liquid is heated, the random movements of its molecules become more rapid as the temperature rises. However, since many of the molecules in liquid water are held firmly together by hydrogen bonds, a large amount of heat is required to overcome this intermolecular bonding and allow the molecules to move more rapidly. Consequently, the *specific heat* of liquid water (the heat required to raise the temperature of 1 g of the substance by 1°C is very high compared with less structured liquids (Table 12.1).

Table 12.1. The specific heats of selected molecular substances.

	Molecular wt	*Specific heat*
Water H_2O	18	4·2 $Jg^{-1}{}^{\circ}C^{-1}$
Ethanol C_2H_5OH	46	2·2
Acetone CH_3COCH_3	58	2·1
Carbon tetrachloride CCl_4	154	0·8

Similarly, when ice melts, or liquid water evaporates, large quantities of heat energy are needed to break the hydrogen-bonded structure so that the molecules can escape and move more freely. As a result, the temperature of these changes of state are unusually high, compared with other molecular substances, as are the Latent Heats of Melting and Vaporization (Table 12.2) ; where the *Latent Heat of Melting* of a substance is the heat required to change 1 g of the substance from the solid state at its melting point to the liquid state at the same temperature ; similarly for the *Latent Heat of Vaporization*, from liquid to gaseous state.

Table 12.2. Thermal properties of selected molecular substances.

	m.p. (°C)	*b.p. (°C)*	*Latent heat of melting* (Jg^{-1})	*Latent heat of vaporization* (Jg^{-1})
Water	0	100	334	2260
Ammonia	—78	—33	332	1374
Ethanol	—117	79	105	854
Acetone	—95	56	98	521
Carbon tetrachloride	—23	77	18	194

The influence of hydrogen bonding on changes of state is clearly illustrated by the melting and boiling points of the hydrides of Group 6 elements, i.e. H_2O, H_2S, H_2Se and H_2Te. If, as we might at first expect, the intermolecular bonding of these substances were similar, then we should predict that H_2O (the heaviest) would have the highest values. (Less energy should be required to accelerate a lighter particle.) However, as shown in Table 12.3, the relationship between molecular weight and the temperatures of melting and boiling holds only for H_2S, H_2Se and H_2Te. In H_2O, the intermolecular hydrogen bonds are so much stronger than in the other three hydrides that it has the highest m.p. and b.p. rather than the lowest. (It is worth stressing that according to its molecular weight, water should be a gas at room temperature.)

Table 12.3 Melting and boiling points for the hydrides of Group 6 elements.

Hydride	*Molecular wt*	*m.p. °C*	*b.p. °C*
H_2O	18	0	100
H_2S	34	—86	—61
H_2Se	81	—60	—42
H_2Te	130	—49	—2

Because of its high specific heat, water is very good temperature 'buffer', absorbing or losing large amounts of heat without excessive variation in temperature. This is particularly important in warm-blooded animals whose body temperatures must be maintained close to an optimum value. It is also important in controlling the temperature of soils ; the surface layers of a bare dry soil can experience enormous daily fluctuations of temperature (variations of 30—40°C are possible), causing considerable damage to plants. In moist soils, the high specific heat of water causes such fluctuations to be 'damped', giving lower maximum and higher minimum temperatures. In general, the high specific heat of water has a moderating effect on the temperatures of plants and animals, soils, lakes, the atmosphere, etc.

The high latent heats of water are also important in the control of temperature. In particular, both plants and animals remove excess heat from their surfaces by the evaporation of water. For each gram of water lost to the atmosphere by perspiration or transpiration (at 25°C), 2441 J of heat is also lost from the skin or leaf surface. On the other hand, the high latent heat of vaporization of water accounts for the serious burns or scalds caused by the exposure of skin to steam ; here the condensation of each gram of steam releases 2260 J of heat (at 100°C).

In temperate agriculture, the damage to crop plants caused by severed frost can often be avoided by spraying the crop with water. As the applied water begins to freeze on the foliage of the crop, the latent heat released (334 J g^{-1}) maintains the temperature of the plant tissues above 0°C.

OTHER PROPERTIES OF WATER

Water as a Solvent

Water is an excellent solvent for three groups of biologically important solutes :

(a) *Organic solutes with which water can form hydrogen bonds*, including amino acids and low molecular weight carbohydrates, and proteins, which contain hydroxyl, amine or carboxylic acid functional groups. Water also forms colloidal dispersions with larger carbohydrate and protein molecules (e.g. the cytoplasm). Similarly, water 'wets', or adheres to, solid substances such as glass, cotton and clay mineral which carry hydroxyl groups on their surfaces.

(b) *Charged ions* such as ionic salts dissolve readily because (partially charged) water molecules orientate themselves round ions in the crystal and 'pull' them into solution as highly soluble, hydrated ions. In the same way, water molecules become attached to fixed charges on the surfaces of plant cell walls, cell membranes and soil particles, giving tightly bound layers of water a few molecules thick.

(c) *Small molecules*, like the atmospheric gases, which can fit into holes in the open, cage-like, structure of liquid water.

Thus as well as being an ideal solvent for biochemical reactions, water is also a suitable medium for the transport of organic molecules (e.g.

sucrose in blood and phloem), nutrient ions (e.g. nutrients from root to leaf in the xylem) and atmospheric gases (e.g. movement of O_2 to site of respiration).

Substances which dissolve in, or bind water (groups (a) and (b)) are termed *hydrophilic*, i.e. water loving, whereas those which neither dissolve in, nor bind water are *hydrophobic*, water hating. Hydrophobic substances carry neither electrical charges nor groups forming hydrogen bonds with water; they include hydrocarbons, halogenated hydrocarbons, fats, oils, greases, etc. Many organic molecules are *amphipathic*, containing both hydrophobic and hydrophilic groups.

Volume Changes in Ice and Water

In general, when solids are heated, they expand, and this expansion continues throughout the solid and liquid states. However, as shown in Fig 12.2, water undergoes an unexpected *contraction* around its melting pint; as would be predicted, ice expands with rising temperature but, at its melting point, there is a large and sudden contraction in volume (≃ 9%) followed by a more gradual contraction between 0 and 4°C, at which temperature liquid water achieves its highest density (Figure 12.2). Above 4°C, the liquid reverts to a more conventional steady expansion with increasing temperature.

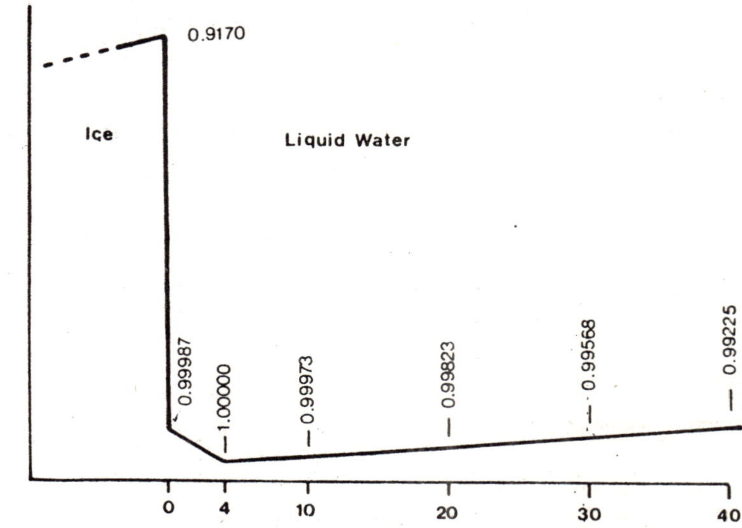

Fig. 12.2 Changes in the volume of a fixed mass of water between 0 and 40°C, (not to scale). The density of water (or ice) is given at ten-degree intervals and at 4°C, the temperature of maximum density and minimum volume.

The reason for this unusual behaviour is that the crystal structure of ice is a rather open, cage-like lattice (as in the 'icebergs' of liquid water). When the crystal structure breaks down on melting, individual water molecules are more mobile and can approach one another more closely, thereby occupying a smaller volume.

Consequently, when a body of water (puddle, pond, lake) cools from 10°C, the cooler (denser) water falls to the bottom until the temperature of all the water has fallen to 4°C. Water cooler than this critical temperature is less dense and remains at the surface, where ice forms first. This phenomenon, which is of the greatest importance for the survival of aquatic organisms in cold climates, is a clear example of the unique relationship between the properties of water and the needs of living organisms.

Tensile Strength and Viscosity

Two other properties, tensile strength and viscosity, are highly important in the movement of water in animals, plants and soils. In particular, the unusually high tensile strength, or cohesion, or water columns in the xylem (due again to intermolecular hydrogen bonding) means that water can be drawn to the tops of tall trees by transpirational pull alone.

Table 12.4 The viscosity of a range of liquid substances.

	Viscosity (centipoises)	*Temperature (°C)*
Acetone	0.316	25
Carbon tetrachloride	0.969	20
Water	1.002	20
Ethanol	1.200	20
Glycerol	1.490	20
Mercury	1.554	20
Light machine oil	133.8	15
Heavy machine oil	660.6	15

The viscosity of a liquid is a measure of its ''stickiness' and, therefore, it controls the rate at which the liquid will pour from a spout or flow along a pipe. The viscosity of a liquid can be measured by timing the fall of a metal sphere through a column of the liquid ; familiar liquids of high viscosity include treacle, honey and thick lubricating oils. In spite of its strong intermolecular hydrogen bonds, water has a moderate viscosity (when compared with other molecular liquids (Table 12.4)), permitting rapid mass flow through pipes (irrigation pipes, blood vessels, soil pores, xylem conduits, etc.) in response to modest pressure gradients. The rapid fall in the viscosity of water with rising temperature (Table 12.5) also has practical consequences ; for example, tropical soils drain more rapidly than temperature.

Table 12.5 The influence of temperature on the viscosity of liquid water.

Temperature °C	0	5	10	15	20	25	30	35
Viscosity (centipoises)	1·787	1·516	1·306	1·138	1·002	0·890	0·798	0·719

Other Properties

The transparency of liquid water ensures a supply of light for photosynthesis to submerged plants and algae ; similarly the transmission of light through cytoplasm means that energy for photosynthesis can be captured by several layers of cells in the mesophyll of the leaves of terrestrial plants. Other important properties of water include its *dissociation into ions* and its unusually high *surface tension*.

THE PURIFICATION OF WATER

Water has many uses and for each use it must achieve certain standards of purity. For example, it is necessary for irrigation water to contain low concentrations of salts so as to avoid the build up of soil salinity under conditions of rapid evaporation.

For industrial and household (washing) purposes, 'soft' water, with a low dissolved salt content, is preferred. ''hard' water, containing significant concentrations of Ca, Mg and Fe salts (particularly bicarbonates and carbonates) normally originates from limestone catchments or from boreholes receiving large quantities of leached salts. Over a period of time, calcium salts from hard water accumulate as deposits of 'fur' or 'scale' in boiler pipes and in kettles, reducing water flow and decreasing the efficiency of heating. The salts in hard water also cause the precipitation and inactivation of soaps and detergents. These problems can be overcome by passing the water through an ion exchanger or by the use of special synthetic detergents.

Rigorous controls must be exercised over the quality of piped drinking water which must be free of suspended solids, harmful microorganisms and toxic chemicals. Normally, most of the solids and microorganisms are removed by passing the water through sedimentation tanks and slow filter beds ; the remaining organisms can then be killed by the addition of chlorine or another sterilizing agent. The exclusion of toxic chemicals can be achieved only by firm controls over the catchment area, e.g. ensuring pesticides do not drain directly into reservoirs, and by regular analysis of the purified drinking water.

In the laboratory, we require pure water containing very low concentrations of all impurities. For most purposes, water distilled using glass apparatus is sufficiently pure, but even this contains, amongst other impurities, alkali metal ions dissolved out of the glass. Ultrapure water for analytical work can be obtained by distillation using tin equipment followed by deionization, i.e. the removal of both cations and anions using an ion exchange resin.

However, it must be stressed that even ultrapure distilled and deionized water must contain hydronium and hydroxide ions (each at 10^{-7} M) according to the relationship :

$$K_w = [H_3O^+]\,[OH^-] = 10^{-14}.$$

In addition, distilled water rapidly absorbs gaseous impurities (CO_2 O_2, N_2, etc.) when left open to the atmosphere.

ION EXCHANGE IN SOILS AND IN THE PURIFICATION OF WATER

Ion Exchange in Soil

Clay and humus particles carry, on their surfaces, negative charges which can bind cations, particularly H^+, NH_4^+, $^*Al^{3+}$, Ca^{2+}, Mg^{2+}, and K^+. These bond cations are important in determining the pH of the soil as well as in supplying the growing plant with nutrients. A fertile soil may have 65% of these negative charges (cation exchange sites) occupied by Ca^{2+}, 10% by Mg^{2+}, 5% by K^+ and 20% by H^+. However, this is not a static situation ; cations in the soil solution can replace bound cations and this replacement is called *cation exchange.* Cation exchange *in soils* is governed by two rules :

(a) The 'replacing power' of cations is as follows :

$$H^+ > {}^*Al^{3+} > Ca^{2+} > Mg^{2+} > K^+ > Na^+$$

For example, at *equal concentrations* K^+ ions will replace Ca^{2+} ions on the exchange sites more completely than will Na^+ ions, and so on. Overall, H^+ (H_3O^+) and $^*Al^{3+}$ ions are the most efficient at dislodging cations. This order of cations is also the order of strength of binding to the exchange sites, i.e. H^+ will be more firmly bond than $^*Al^{3+}$ etc. *

---

* Al^{3+} ions take part in ion exchange only at low pH

(b) Law of Mass Action. The 'replacing power' of an ion depends also upon its *concentration*, in the soil solution. Thus Na^+ at a high concentration can replace ca^{2+} more completely than K^+ at a lower concentration.

These rules may be illustrated by considering what will happen to our fertile soil (see above) under cropping. Uptake of calcium by the roots of the crop plants will tend to lower the soil solution concentration of Ca^{2+} such that, by rules (a) and (b), H_3O^+ ions in the soil solution will begin to replace Ca^{2+} ions on the exchange sites (III) thereby releasing Ca^{2+} ions into the soil solution where they can, in turn, be lost from the soil by plant uptake or by leaching. This process can therefore continue until a large proportion of the exchange sites are occupied by H^+ ions. At first sight it would seem that, according to rule (a), these sites should now be permanently occupied by H^+ ions (the most firmly bound ions), giving a permanently acidic and infertile soil. However, we must not ignore rule (b). When we apply lime ($CaCO_3$ or $Ca(OH)_2$) to the soil we increase greatly the soil solution concentration of Ca^{2+} ions which, by rule (b), can now replace H^+ ions on the exchange sites. Similar phenomena occur when potassium and ammonium fertilizers are applied to soils.

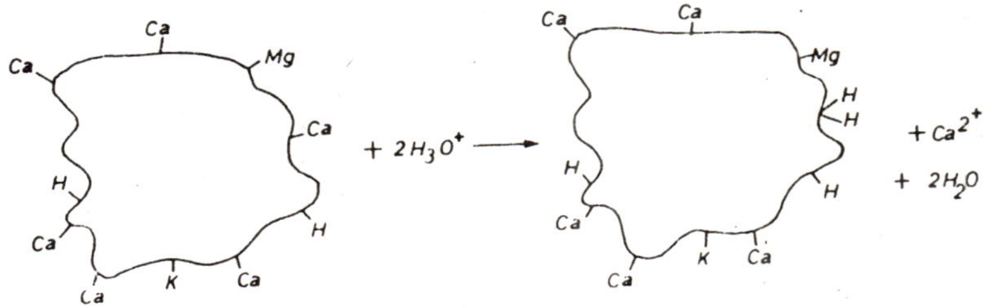

The Deionization of Water

The purification of water by ion exchange resins proceeds in two stages. First, the water is passed through a column containing a polymer (resin) carrying cation exchange sites saturated with H^+ ions. These cation exchange sites have a higher affinity for metallic cations than for hydrogen (compare with soil, rule (a)), and, therefore, reactions like :

$$\boxed{\text{Resin}}\text{—H} + Na^+ + H_2O \rightarrow \boxed{\text{Resin}}\text{—Na} + H_3O^+$$

remove all the cationic impurities and release hydronium ions. The water is now passed through a second column containing a resin carrying anion exchange sites saturated with hydroxide ions. Reactions like :

$$\boxed{\text{Resin}}\text{—OH} + \text{Cl}^- \rightarrow \boxed{\text{Resin}}\text{—Cl} + \text{OH}^-$$

remove anions from solution and release hydroxide ions.

All anion and cation impurities have now been removed and the following neutralization reaction occurs :

$$H_3O^+ + OH^- \rightarrow 2H_2O$$

Once all the exchange sites on the resin have become saturated with impurities, the columns can be regenerated using concentrated acid and alkaline solutions (according to rule (b)).

In general, the behaviour of an ion-exchanging material (soil, resin, etc.) towards a solution containing ionic solutes can be described by a replacement/affinity/binding series, as in rule (a), which is characteristic of the ion-exchanging material, modified by the Law of Mass Action.

CHAPTER 13

Surface and Colloid Chemistry

Several crops (e.g. cotton) must be sprayed very frequently with insecticide to prevent loss of yield and deterioration in the quality of the product. However, if we examine a cotton plant during rain, it is clear that raindrops do not adhere to the surfaces of the leaves but roll off almost immediately. How can it be, then, that the droplets of an insecticide spray do adhere to the leaf and spread out to give an even distribution of the dissolved chemical over its surface ? In this chapter we shall try to answer this question, and also to explain some features of the surfaces of soil particles, by studying the physical chemistry of liquid/solid interfaces.

SURFACE TENSION

In the bulk of any liquid, each molecule is attracted equally in all directions to the surrounding molecules. However, at the surface, the outermost molecules tend to be pulled inwards (into the bulk of the liquid) because there is no molecular attraction from outside the liquid surface—

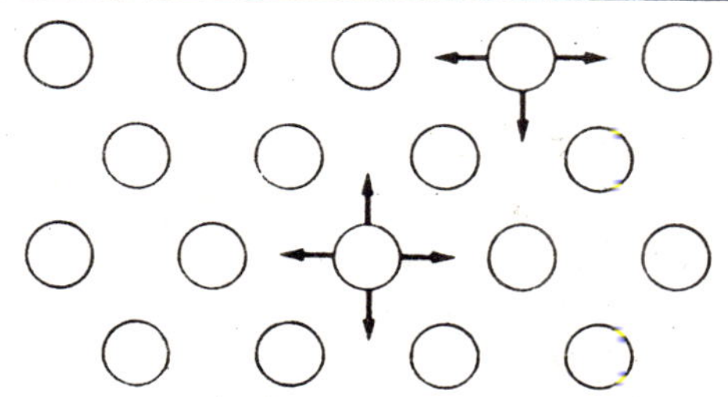

These inwardly directed forces cause small volumes of the liquid to assume the shape with the lowest surface area : volume ratio. This shape is a sphere—

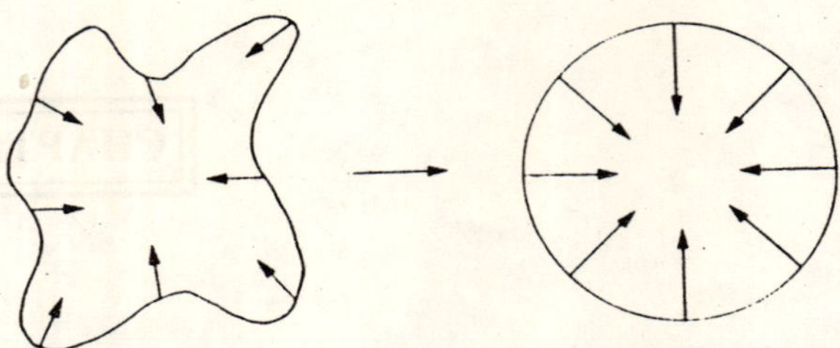

Once a volume of liquid has taken up a spherical shape, these inwardly directed forces resist any expansion in surface area. If we do wish to increase the surface area (e.g. to spread the droplet out as a thin layer covering a leaf), then work must be done against these forces, and the stronger the forces are, the greater the amount of work required to give unit increase in surface area.

The Surface Tension of a liquid is an index of the amount of work required to give unit increase in the surface area of a volume of liquid (measured under standard conditions). As shown in Table 13.1, liquids with unusually strong intermolecular forces (e.g. water) or interatomic forces (mercury) have high surface tensions.

Table 13.1. The surface tension of selected liquids in contact with air at 20°C.

	Surface tension mN m^{-1}
Water	72.75
Benzene	28.83
Acetic acid	27.60
Carbon tetrachloride	26.80
Acetone	23.70
Ethanol	22.30
Mercury	485.00

THE WETTING OF SURFACES—DETERGENTS

Up to this point, we have considered liquid droplets in air, but if we now place a water droplet on a hydrophilic surface (e.g. glass or cotton cloth) which carries hydroxyl groups.

then hydrogen bonds can form between the outermost molecules in the water droplet and the hydroxyl groups on the hydrophilic surface. This 'neutralizes' the inwardly directed force, reduces the surface tension of the water and enables the droplet to spread thinly over the surface. Thus the water in the droplet can *wet*, and adhere to, the surface.

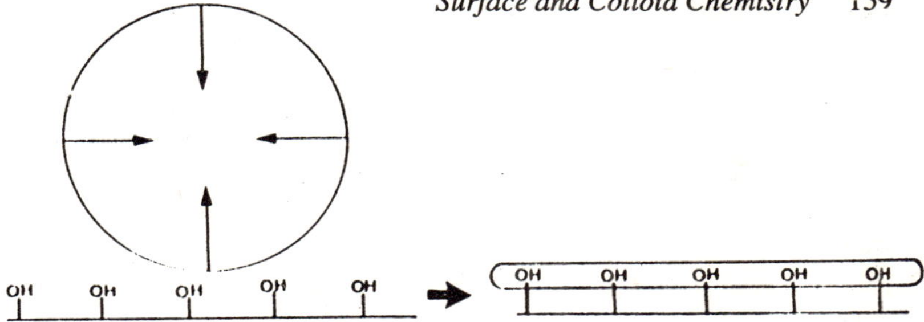

On the other hand, if we place the water droplet on a hydrophobic surface with which it cannot form bonds, then the high surface tension of water results in *no wetting* of the surface. There is a minimum of contact between liquid and surface, and the droplet will roll off.

The successful wetting of hydrophobic surfaces is an important problem in the application of pesticides. In spraying a crop with insecticide or fungicide, it is necessary to cover the leaves uniformly will spray so that when the water evaporates, the active ingredient is uniformly distributed, giving reasonably complete protection. However, leaf surface are hydrophobic, due to a thin coating of wax over the epidermis, and, therefore, aqueous solutions and suspensions roll off without wetting the surface. Similarly, the hair and hides of livestock tend to be greasy and hydrophobic, repelling aqueous dip solutions.

It is not feasible to solve this problem by modifying the hydrophobic surfaces of crop plants and livestock animals, because they are important in protecting the epidermis against damage, desiccation or heat loss. Instead it is necessary to alter the surfaces of spray droplets so that they adhere to hydrophobic surfaces ; this is achieved using detergents.

Detergents are amphipathic organic molecules containing :

(a) a hydrophilic functional group, attached to

(b) a long hydrophobic hydrocarbon chain (normally longer than 8 carbon atoms).

For example *soaps*, the sodium salts of long-chain carboxylic acids, are the most familiar detergent substances, e.g. sodium stearate :

$CH_3—CH_2—CH_2—CH_2—CH_2—CH_2—CH_2—CH_2—CH_2—CH_2—$
$CH_2—CH_2—CH_2—CH_2—CH_2—CH_2—CH_2—CO_2—Na^+$

or more concisely, $Ch_3(CH_2)_{16}—CO_2—Na^+$

and in diagram form—

CO_2 Na^+

Synthetic Detergents are very similar to soaps but normally contain a sulphate (—O—SO_3 –), sulphonate (—$C_6H_4SO_3$ –) or, phosphate group instead of a carboxylate.

Although their molecules are predominantly hydrophobic, soaps and synthetic detergents dissolve in water due to the clustering of their molecules into *micelles*—

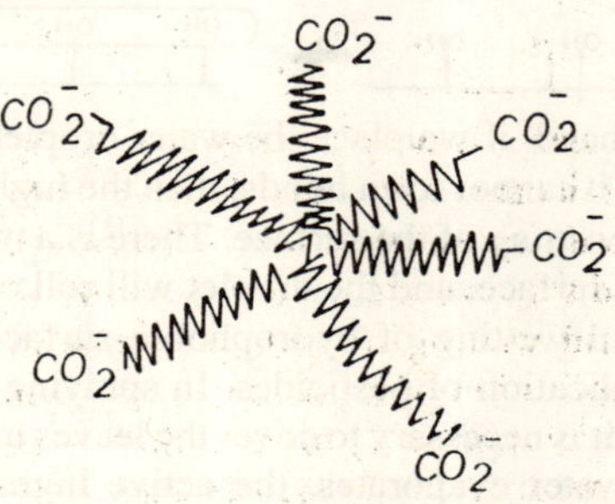

By this arrangement, all the hydrophobic chains are 'hidden' from the solvent, whereas all the hydrophilic groups are in contact with, and can form bonds with, water molecules. Thus, a detergent micelle can be though of as a giant ion carrying many negative charges ; in addition to these hydrophilic interactions between micelle and water, the mutual repulsion of the negatively charged micelles also favours the solubility of detergents.

When a crop spray droplet containing a detergent (or 'spreader') comes into contact with a leaf, these micelles break up because the hydrophobic 'tails' of the detergent molecules are attracted to the hydrophobic surface of the leaf. However, because the hydrophilic ends of the detergent molecules remain firmly 'anchored' in the water of the droplet, the detergent molecules form a 'bridge' between the aqueous solution and the leaf surface, thereby overcoming the high surface tension of water—

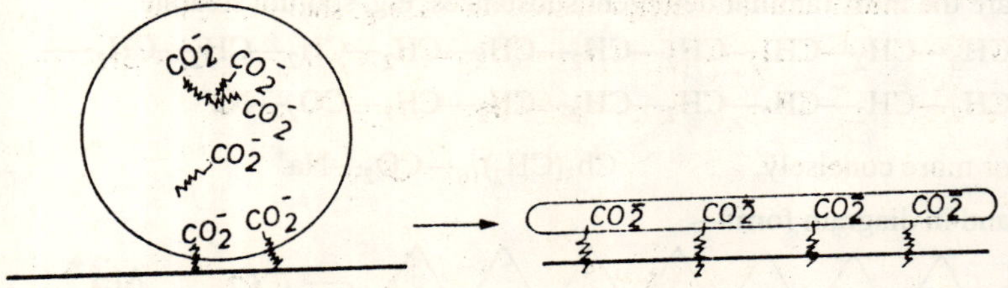

Consequently, the spray droplets adhere to, and cover, the leaf surface.

A more familiar example of detergent action is the removal of dirt from the hydrophobic surfaces of skin using soap and water. Water alone is not effective in cleaning the skin because of repulsion by the hydrophobic ends remain in the water. When we rub our hands in soapy water, particles of dirt are detached from the skin at the centre of soluble micelles—

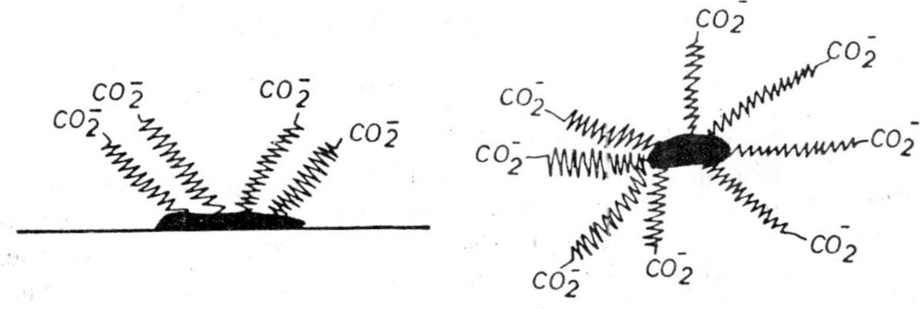

Soaps cannot be used in 'hard' water because calcium, magnesium and iron cations cause the precipitation and inactivation of the soap anions, e.g.

$$Ca^2 + 2CH_3(CH_2)_{16}CO_2^- \rightarrow Ca(CH_3(CH_2)_{16}CO_2)_2 \downarrow .$$

calcium stearate

Instead, a synthetic detergent which does not precipitate may be used. Detergents have been devised for a wide range of applications including the dispersion of oil pollution in the sea.

INTERACTIONS BETWEEN SOIL SURFACES AND WATER

Soils consist of mineral particles (mostly hydrophilic) of varying diameter (sand 2 0·06 mm, silt 0·06 0·002 mm and clay < 0·002 mm. Soil Survey of England and Wales classification), bound together into aggregates by organic matter and clay particles. Within and between these aggregates, there is a network of interconnected pores of diameter ranging from *a few cm* (drying crack, earthworm or termite channels), through *a few mm* (pores between aggregates), down to *a few μm or tenths of a μm* (finest pores within aggregates). Thus the internal surface area : volume ratio of a well-aggregates). Thus the internal surface area : volume ratio of a well-aggregated soil is enormous.

When a dry soil is first wetted, water is attracted to the hydrophilic surfaces of soil particles and tends to be spread thinly over a very large surface area. However, as we have seen in section 10.1, water tends to change its shape so its to *minimize* its surface area in contact with air. In soils, this reduction in the air/water interface is achieved by the *filling* of pores with water ; once a soil pore has been filled, work must be done against the surface tension of water to withdraw water from the pore, because withdrawal involves an increase in the water surface in contact with air.

For example, after the drainage of a temperate soil is complete, all soil pores narrower than 60 µm (10 µm in tropics) will be filled with water, bound by the hydrophilic pore walls and by the surface tension of water. As pores become narrower, these retaining forces (normally called the *Matric* forces) increase as shown by the expression :

$$\text{Suction required to withdraw water from pore (bars)} = \frac{3}{d}$$

where d is the pore diameter in µm.

Another feature of soil/water relationship originating from surface properties is the *capillary movement* of water. This phenomenon can be explained using the simpler glass/water system. For example, if we insert the end of a wide bore (several mm in diameter) glass tube into a beaker of water, water molecules are attracted to, and pulled up, the hydrophilic internal surface of the tube, giving the familiar concave, rather than flat, meniscus. However, there is no upward movement of the bulk of the water inside the tube—

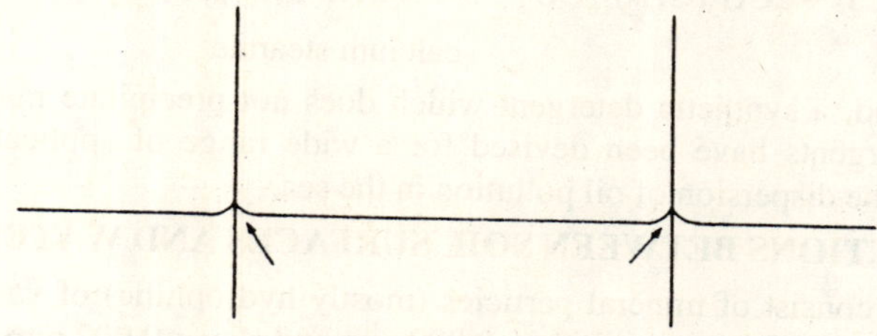

If we reduce the internal diameter of the glass tube, then the *proportion* of water molecules in contact with 'the internal surface of the tube

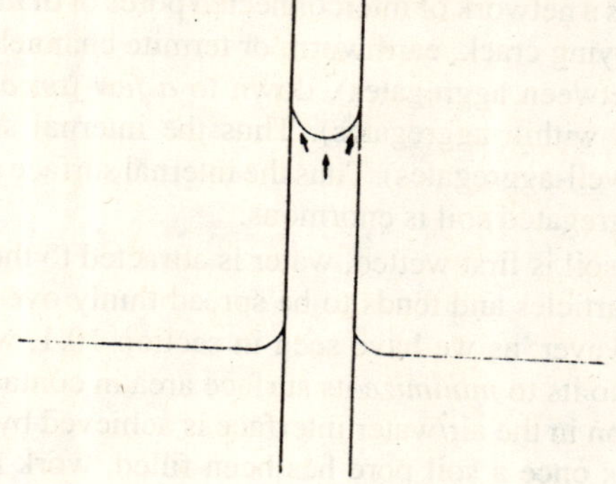

increases whereas the mass of the column of water in the tube decreases. Consequently, at a certain diameter, the upward-directed forces between glass and water become large enough to raise the water column.

For example, an internal diameter of 1 mm gives capillary rise of 30 mm above the water level in the beaker compared with 1·5 m with a diameter of 0·02 mm. Such large rises are possible only because the strong intermolecular forces in water allow transmission of the force from the molecules in contact with the glass to the bulk of the liquid.

Capillary rise, or capillarity, is responsible for much of the vertical and lateral movement of water in soils. The role of capillary action in bringing water to the soil surface from a deep water table has been the subject of considerable controversy, but it can be demonstrated in the field that capillary rise can subirrigate crops adequately from a water table at a depth of at least 1 metre.

THE COLLOIDAL STATE

If we take a cube of material of side 1 m and cut it into 8 identical cubes, each of side $\frac{1}{2}$ m, then the total mass and volume of material remains the same but the surface area has increased from 6 m^2 to 12 m^2. If each of these smaller cubes is similarly divided, we obtain 64 cubes of total surface area 24 m^2. Thus the surface area : volume ratio is doubled each time the side is halved (Table 13.2). Consequently, if a substance is divided up very finely, its surface area per unit volume (or mass) becomes very large and the properties of its surface come to dominate its chemical behaviour. For example many powders, such as finely ground flour, can burn spontaneously in air although the bulk material requires activation energy to be supplied before combustion can take place. Similarly, the behaviour of a *colloid* tends to be determined more by its surface area than by the chemical properties of the material in bulk.

Table 13.2 The effect of subdivision of a solid cube on surface area.

Cube side	*No. of cubes*	*Total volume* (m^3)	*Total surface area* (m^2)	*Surface area volume*
1 m	1	1	6	6
$\frac{1}{2}$	8	1	12	12
$\frac{1}{4}$	64	1	24	24
$\frac{1}{8}$	512	1	48	48
and so on.				

Matter exists in the *Colloidal State* when it is divided up into small particles of diameter 10^{-6} to 10^{-9} m ; thus colloids are defined by the size of their particles, which may consist of very large molecules (macromolecules) or clusters of molecules of atoms.

Colloidal Systems consist of matter in the colloidal state (the *Dispersed Phase*) spread uniformly throughout a *Dispersion Medium.* In theory, both dispersed phase and dispersion medium can exist in each of the three states of matter, but the more familiar combinations are shown in Table 13.3. In the following section, we shall concentrate on sols and emulsions which are the most important kinds of colloidal system in environmental chemistry.

Table 13.3 Some familiar colloidal systems.

Dispersed phase	*Dispersed medium*	*Colloidal system*	*Example*
Liquid	Gas	Liquid aerosol	Fog
Solid	Gas	Solid aerosol	Smoke, dust
Gas	Liquid	Foam	Soap lather, fire extinguisher foam
Liquid	Liquid	Emulsion	Milk
Solid	Liquid	Sol, paste	Toothpaste

COLLOIDAL SYSTEMS

It is useful to recognize three distinct classes of sols (and emulsions).

(a) *True solutions of macromolecules.* The most familiar examples of this type of sol are solutions of hydrophilic macromolecules (proteins, carbohydrates) in water. In particular, the cytoplasm is a complex solution of many different macromolecules, predominantly proteins. Where the dispersion medium is other than water (e.g. rubber in benzene), the dispersed phase is called *Lyophilic*, solvent loving.

As well as possessing many of the normal properties of solutions, solutions of macromolecules have other characteristics ; in particular :

High Viscosity. This accounts for the name colloid which means glue-like.

Dialysis. Unlike smaller solutes, macromolecules in solution do not pass through the pores of dialysis membranes. Consequently, dialysis is a useful technique in biochemistry for separating macromolecules from other, usually inorganic, solutes.

(b) *Association Colloid Systems* are formed when a number of amphipathic molecules (containing lyophobic and lyophilic groups) cluster together to form *micelles* of colloidal dimensions. Soaps and detergents form association colloid systems in water.

(c) *Colloidal Dispersions.* Here the dispersed phase consists of atoms or molecules clumped together into particles of colloidal size. These are not true solutions because the dispersed phase is normally not strongly lyophilic ; however, the clumps are not forced together by the solvent to give a precipitate because they tend to accumulate (charged) ions from solution on their surfaces. As a result of these charges, the clumps repel one another strongly and remain in suspension.

The hydrated iron oxide/water sol is an easily prepared colloidal dispersion, often demonstrated in the laboratory. It can be prepared by boiling a ferric chloride solution to give hydrated iron oxide $Fe_2O_3.xH_2O$ which is insoluble in water. However, as the molecules of iron oxide begin to clump together, they absorb hydronium ions from solution, giving a positive surface charge—which repels other positively charged colloidal particles ; the dispersion is, therefore, stabilized by the repulsion of neighbouring colloid particles. The addition of large anions like the ferricyanide ion ($Fe(CN)_6^{3-}$) causes the iron oxide to precipitate (or *flocculate*) by neutralizing the positive charges—

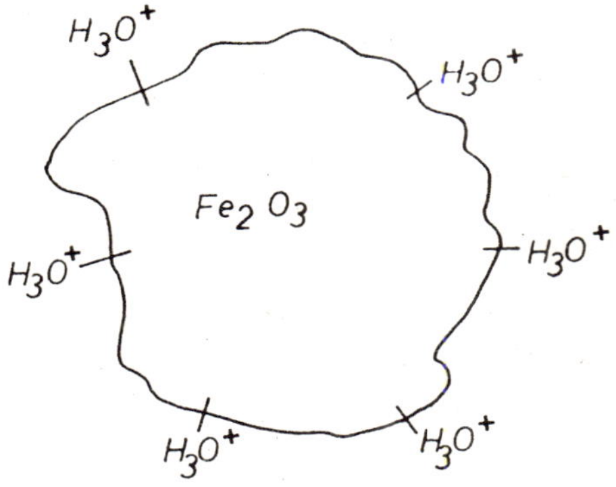

Silicate Clays

Silicate clays are similar in structure to sheet silicates but with aluminium hydroxide layers interposed between the silicate sheet. However, in contrast to the large flaky crystals of mica, silicate clays in soils exist predominantly in the colloidal state. in addition, these colloidal particles are negatively charged, due to the isemorphous replacement of Al for Si and Mg for Al. some important properties of these common clay minerals are given in Table 13.4.

Soil clay particles, therefore, have enormous surface areas upon which occur many of the important reactions of soil chemistry. In particular, these surfaces (together with the surfaces of humus particles, also

Table 13.4. Selected properties of silicate clay minerals in soils.

Clay mineral	*Particle size*	*surface area*	*Negative charge*
Kaolinite	0·1 –5·0 μm	5—20 $m^2 g^{-1}$	3—15 mEq* 100 g^{-1}
Illite	0·1 –2·0	100—120	15—40
Montmorillonite	0·01 – 1·0	700—800	80—100

*Using the ion-exchange definition of an equivalent.

colloids) are the sites of cation exchange and of deposition of the products of the weathering of soil minerals (e.g. iron oxides).

It is essential for the particles (sand, silt and clay) of agricultural soils to bind together into stable aggregates, since this promotes drainage and aeration and increases the amount of water available to plant roots. The primary process in the formation of aggregates is the flocculation of colloidal clay particles into *domains* which, with humic material, bind the sand and silt particles together.

As with micelles and colloidal dispersions, clay particles in water tend to remain dispersed because of mutual repulsion of their negatively charged surfaces. However, in soil these negative charges are neutralized by cations. The idea of cation exchange *sites* is useful when discussing cation exchange but when considering the flocculation of clay, it is more helpful to consider the negative charge to be spread over the whole colloid particle, and shielded from neighbouring particles by a 'cloud' of cations. (The negative charge at the colloid surface and its accompanying cloud of cations are together normally called the *diffuse double layer.*) Cations differ in their ability to shield the negative charges on colloid particles according to the series :

$$Al^{3+} > Ca^{2+}, H^+ > Mg^{2+} > K^+ > Na^+$$

For example, Ca^{2+} ions are so successful in shielding the negative charges on adjacent clay particles that the particles cease to repel one another and flocculation can occur. On the other hand Na^+ ions are much less successful, causing the clay to remain dispersed or deflocculated.

Consequently, soils with a high Ca^{2+} ion content, maintained by liming, tend to have good physical properties due to the aggregation of soil particles. In contrast, in saline soils, the deflocculation of clay leads to poor aggregation and unfavourable physical conditions crop growth.

PART II
ORGANIC CHEMISTRY

CHAPTER 14

Introduction to Organic Chemistry

THE UNIQUENESS OF CARBON

Organic chemistry is the study of covalently bonded carbon compounds, although ionic bonds do occur in a few organic substances. One of the reasons for studying carbon compounds separately from those of the reasons for studying carbon compounds separately from those of the other 103 (or more) elements is simply that there are more compounds of carbon than of all the other elements put together. It has been calculated that there are 3 million known organic compounds compared with about 60 000 inorganic compounds. A more important reason is that all living organisms are made up principally of organic substances, and use a vast array of carbon compounds in their growth and physiology.

Carbon forms more compounds than any other element due to a unique combination of three atomic properties.

1. Carbon has a high valency, 4. To obtain a stable octet, each carbon atom can share a pair of electrons with four other atoms, giving four single bonds, for example, in methane, CH_4

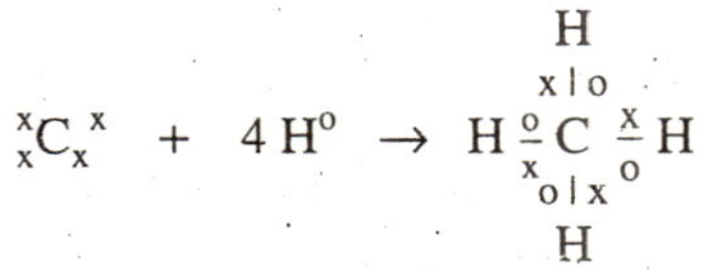

$$
{}^{x}_{x}C^{x}_{x} \;+\; 4\,H^{o} \;\rightarrow\; \begin{array}{ccccc} & & H & & \\ & & {\scriptstyle x \mid o} & & \\ H & \overset{o}{\underset{x}{-}} & C & \overset{x}{\underset{o}{-}} & H \\ & & {\scriptstyle o \mid x} & & \\ & & H & & \end{array}
$$

(Note that although the bounds in singly bounded carbon atoms are directed towards the corners of a tetrahedron, it is normally more convenient in organic chemistry to draw them in one plane.) If carbon forms bonds with a number of different atoms, its high valency alone can result in a large number of possible compounds.

2. Single covalent bonds linking carbon atoms are strong not only in the pure element but also in compounds of carbon with other elements, especially hydrogen. This permits the formation of chains of carbon atoms, both straight and branched and of unlimited length, as well as the formation of rings of carbon atoms, for example—

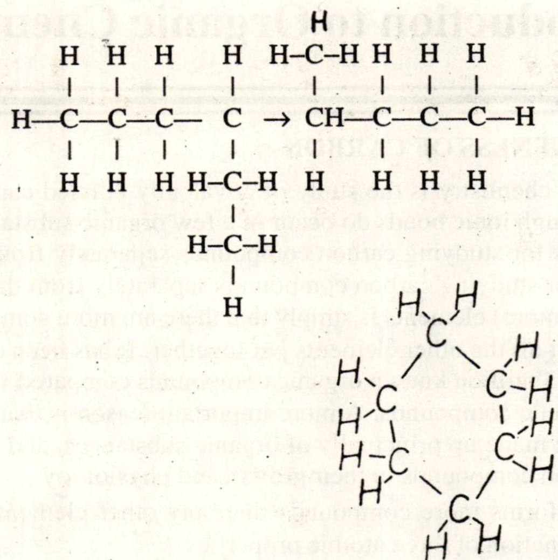

The atoms of several other elements (S,P, etc.) are linked in groups by strong covalent bonds, in the element, but not when their atoms are attached to the atoms of other elements. Because of this linking of carbon atoms into chains and rings of different sizes and shapes, an enormous number of compounds of carbon is possible.

3. Carbon to carbon double and triple bonds are also strong and stable, resulting in a further increase in the variety of possible organic compounds, for example—

```
                                 H
H       H                        |
 \     /                         |
  C = C       H — C ≡ C — C — H
 /     \                         |
H       H                        |
                                 H
```

The atoms of several elements have at least one of these properties but carbon atoms are unique in possessing all three. As a result of A, B and C, we can predict the existence of an enormous number of even the simplest organic compounds—the *hydrocarbons*, which contain carbon and hydrogen atoms only.

HOMOLOGOUS SERIES AND FUNCTIONAL GROUPS

In spite of the large number of known organic compounds, it is not necessary to examine the properties and reactions of each compound individually. Instead, organic compounds can be classified into a few *homologous series* of closely related compounds with similar properties and reactions. For example, the series of compounds—

```
     H            H  H             H  H  H
     |            |  |             |  |  |
  H—C—OH      H—C—C—OH       H—C—C—C—OH
     |            |  |             |  |  |
     H            H  H             H  H  H

            H  H  H  H
            |  |  |  |
         H—C—C—C—C—OH      -------
            |  |  |  |
            H  H  H  H
```

is a homologous series of *alcohols*, each containing the hydroxyl (OH) *functional group* attached to a hydrocarbon chain of varying length. Hydrocarbon chains tend to be chemically inert and, therefore, the chemical properties of the alcohols depend mainly upon the hydroxyl functional group. Consequently, we can establish generalizations about the properties of alcohols by studying a few representative members of the series.

This simplifies the study of organic chemistry considerably, since, for example, a complex molecule containing a hydroxyl group will have at least some of the characteristic properties of the alcohols. Note, however, that although the length and branching of the hydrocarbon chain may have only a small effect upon the chemistry of a molecule, it will have a large effect upon the physical properties, especially boiling point and freezing point, and upon the water solubility. This will be considered in more detail in subsequent chapters.

In general, an organic compound consists of a hydrocarbon framework (chains and/or rings) with one or more functional groups attached. (No functional groups in the case of the hydrocarbon homologous series.) Some common functional groups are :

hydroxyl—OH carboxylic acid—C(=O)—OH

aldehyde—C(=O)—H amine —NH_2

ketone —C(=O)— thiol —SH.

Since the hydrocarbon frame work normally has the same chemical properties whether attached to functional groups or not, we shall begin our study of organic compounds by establishing the properties and reactions of the four major groups of hydrocarbons before going on to study molecules carrying functional groups.

CHAPTER 15

The Hydrocarbons

The hydrocarbons, which contain C and H atoms only, may be sub-divided into four homologous series :

(1) The Alkanes (or Paraffins)
(2) The Alkenes (or Olefins)
(3) The Alkynes (or Acetylenes)
} together called the Aliphatic Hydrocarbons

(4) The Arenes (or Aromatic Hydrocarbons)

according to the nature of their carbon to carbon bonds, as described in the following sections.

THE ALKANES

The alkanes are the simplest organic compounds since they are *saturated* (contain no double or triple bonds). Their general formula is C_nH_{2n+2}, for example—

```
                     H
                     |
Methane, CH4     H—C—H
                     |
                     H

                     H   H   H   H   H
                     |   |   |   |   |
n–pentane, C5H12  H—C—C—C—C—C—H
                     |   |   |   |   |
                     H   H   H   H   H
```

Up to this point, we have drawn each organic molecule in full, as for n-pentane above. However, it is much more convenient to use the following condensed forms :

$$CH_3—CH_2—CH_2—CH_2—CH_3 \text{ or } CH_3\ CH_2\ CH_2\ CH_2\ CH_3$$

Some other members of the alkane homologous series are given in Table 15.1.

The Naming of Alkanes : Structural Isomers

The first four members of the alkane series, methane (1C), ethane (2C), propane (3C) and butane (4C) have special names. The remaining members are named by adding the suffix -ane to the Greek word for the number of carbon atoms in the hydrocarbon chain. Thus we have :

pent	-ane (5C)
hex	-ane (6C)
hept	-ane (7C)
oct	-ane (8C)
non	-ane (9C)
dec	-ane (10C)
pentadec	-ane (15C) etc.

When an alkane chain is attached to a functional group, we replace the -ane suffix with an -yl suffix,

e.g. $C_2H_5—Cl$ ethyl chloride

$C_6H_{13}—Cl$ hexyl chloride.

Groups like C_2H_5- and $C_6H_{13}-$ are called *alkyl* groups. This naming system is quite adequate for straight-chain alkanes, but as we saw in section 11.1, branched chains can also exist. In the cases of methane, ethane and propane, we can arrange the carbon atoms only in straight chains :

$$CH_4 \quad CH_3—CH_3 \quad CH_3—CH_2—CH_3$$

However, when we come to butane, C_4H_{10}, the carbon atoms can be arranged in two ways :

(a) $CH_3—CH_2—CH_2—CH_3$

(b) $CH_3—\underset{}{\overset{\overset{\displaystyle CH_3}{|}}{CH}}—CH_3.$

(a) and (b) are not identical molecules. They have the same formula C_4H_{10} but different structures and are called *structural isomers* of butane.

In general, structural isomers tend to have similar chemical properties but different physical properties.

We may call (a) n-butane and (b) iso-butane where n (or normal) indicates a straight chain and iso indicates the presence of--

$$\begin{array}{l} CH_3 \diagdown \\ \qquad CH — \\ CH_3 \diagup \end{array}$$

at the end of the chain. However, there exists more systematic method of naming alkanes (and all other organic compounds) which may be explained using the following complex alkane as an example—

$$\begin{array}{ccccccccccccccccccccc} & & H & & H & & H & & C_3H_7 & & H & & CH_3 & & H & & H & & CH_3 & & H & \\ & & | & & | & & | & & | & & | & & | & & | & & | & & | & & | & \\ H & — & C_{10} & — & C_9 & — & C_8 & — & C_7 & — & C_6 & — & C_5 & — & C_4 & — & C_3 & — & C_2 & — & C_1 & — H \\ & & | & & | & & | & & | & & | & & | & & | & & | & & | & & | & \\ & & H & & H & & H & & H & & H & & CH_3 & & C_2H_5 & & H & & H & & H & \end{array}$$

Step 1 : Select the longest hydrocarbon chain in the molecule and name the molecule accordingly. Therefore, the above alkane is a *decane* (10C chain).

Step 2 : Number the C atoms in this chain from 1 up wards.

Step 3 : List the groups attached to the chain in order, indicating to which C atom they are attached, in the following way :

2-methyl-4-ethyl-5, 5-dimethyl-7-propyl-decane

Note that it would be incorrect to use the name :

4-propyl-6, 6-dimethyl-7-ethyl-9-methyl-decane

(numbering from the other end) since the position of branches (and functional groups) must be indicated by the lower set of number (i.e. 2, 4, 5 and 7 rather than 4, 6, and 9).

Using this systematic method, we should call the (a) isomer of butane simply *butane* and the (b) isomer, *2-methyl-propane.*

Properties and Reactions of Alkanes

1. *Physical Properties*

All alkanes tend to have rather similar chemical properties, but their physical properties are more variable. In particular, as shown in

Table 15.1. Properties of some straight-chain alkanes.

Name	*Formula*	*Molecular weight*	*m.p. °C*	*b.p. °C*	*Name of alkyl group*
Methane	CH_4	16	—183	—164	methyl
Ethane	C_2H_6	30	—183	—89	ethyl
Propane	C_3H_8	44	—190	—42	n-propyl
Butane	C_4H_{10}	58	—138	—0·5	n-butyl
Pentane	C_5H_{12}	72	—130	36	n-pentyl
Hexane	C_6H_{14}	86	—95	69	n-hexyl
Heptane	C_7H_{16}	100	—91	98	n-heptyl
Octane	C_8H_{18}	114	—57	126	n-octyl
Decane	$C_{10}H_{22}$	142	—30	174	n-decyl
Octadecane	$C_{18}H_{38}$	254	28	316	n-octadecyl
Eicosane	$C_{20}H_{42}$	282	37	343	n-eicosyl

Table 15.1, their melting and boiling points depend upon the number of carbon atoms in the formula, the values rising as the molecules become larger and heavier. As a result, the first four alkanes (methane to butane) are gases, the next eleven (pentane to pentadecane) are liquids and the remaining members are solids at room temperature.

2. *Hydrophobic Nature*

Alkane (and other hydrocarbon) molecules are non-polar and highly hydrophobic. In the presence of water, they attract one another because they are strongly repelled by polar water molecules. This mutual attraction of hydrocarbon molecules, which is sometimes called hydrophobic bonding accounts for the very low water solubility of alkanes.

The hydrophobic nature of alkane molecules and alkyl groups has many important consequences throughout biology and agriculture. For example, hydrophobic bonding is involved in the behaviour of detergents and in determining the tertiary structures of proteins. At another level, the disastrous consequences for seabirds of oil spills at sea are a result of the attraction between hydrocarbons and the hydrophobic lipids on their feathers.

3. *Combustion*

Because the complete combustion (burning) of an alkane in the presence of adequate oxygen is a highly exergonic reaction, alkanes are useful fuels for heating, lighting and fuelling engines. For example the combustion of methane.

$$CH_4 + 2O_2 \rightarrow CO_2 + 2H_2O,$$

releases $8{\cdot}9 \times 10^5$ J mole^{-1} of methane (213 Keal mole^{-1}).

Other important fuels include :

Natural or Bottled Gas — (for household, laboratory and in ustrial heating) contains Cl to C4 alkanes ;

Gasoline or Petrol — (for internal combustion engines) contains C6 to C9 alkanes ;

Kerosine or Paraffin — (for jet engines and heating) contains C10 to C16 alkanes.

Note that if insufficient oxygen is supplied during combustion, products other than CO_2 and H_2O are formed (e.g. carbon, soot) and less energy is released.

Petrol must not contain too high a proportion of straight-chain alkanes because these tend to cause explosions ('knock') in the cylinders of the engine, giving uneven power output. The 'octane number' of a given petrol supply is a measure of how highly branched the fuel is and is obtained by comparing the performance of the fuel with the highly branched octane 2,2,4-trimethyl pentane ; the higher the octane number, the more even the power output. Most petrol supplies also contain an 'anti-knock' additive, such as tetraethyl lead, $Pb(C_2H_5)_4$ which is the source of significant amounts of lead pollution (air, soil and water) in urban areas and near motorways.

4. *Chemical Unreactivity*

In living tissues, alkanes and alkyl groups are highly unreactive, although under industrial conditions they can undergo a variety of reactions. Consequently, most biochemical reactions involve functional groups rather than the rather inert hydrocarbon framework.

The Occurrence of Alkanes in Nature

The most important source of alkanes is petroleum or crude oil from oil wells in various parts of the world. Petroleum is a complex mixture of many organic compounds which can be divided by distillation into a number of useful fractions, each fraction containing several similar compounds. Elsewhere in nature, alkanes, are uncommon although they do occur as waxes in the hydrophobic cuticles of leaves.

Halogenated Alkanes

The replacement of hydrogen atoms in simple alkanes by halogen atoms (F, Cl, Br, I) gives rise to a number of useful compounds including :

Chloroform (Trichloromethane)	$CHCl_3$	— Anaesthetic
Carbon tetrachloride (Tetrachloromethane)	CCl_4	—Dry cleaning solvent
Methyl bromide	CH_3Br	—Used in soil sterilization
Iodoform (Tri-iodomethane)	CHI_3	—Antiseptic and local anaesthetic
Ethyl chloride	C_2H_5Cl	—Local anaesthetic
γ-BHC (Hexachloro-cyclohexane)	$C_6H_6Cl_6$	—Organochlorine insecticide

Halogenated alkanes, like alkanes, are particularly resistant to chemical attack. Consequently organochlorine pesticides like γ-BHC tend to accumulate in the environment and in fat stores in living tissues.

THE ALKENES

The alkenes are straight- and branched-chain hydrocarbon compounds containing one or more carbon to carbon double bond. They are, therefore, *unsaturated* and their general formula (1 double bond in the molecule) is C_nH_{2n}, e.g.

Ethene (Ethylene) $CH = CH_2$

Propene (propylene) $CH_2 = CH—CH_3$.

Table 15.2 Properties of some straight-chain alkenes.

Name	*Formula*	*Molecular weight*	*m.p. °C*	*b.p. °C*
Ethene	C_2H_4	28	— 169	— 104
Prospene	C_3h_6	42	— 185	— 47
1-butene	C_4H_8	56	—185	— 6
1-pentene	C_5H_{10}	70	— 138	30
1-hexene	C_6H_{12}	84	—140	63
1-decene	$C_{10}H_{20}$	140	—66	171
1-octadecene	$C_{18}H_{36}$	252	18	179
1-eicosene	$C_{20}H_{40}$	280	29	341

Other members of the C_nH_{2n} series are given in Table 15.2. As with the alkanes, the melting points and boiling points of alkenes depend (somewhat irregularly) upon the length of the carbon chain. Note that the insertion of a double bond into a molecule gives a lowering that the insertion of a double bond into a molecule gives a lowering of m.p. and b.p. (for example, compare hexane with 1-hexene in Tables 15.1 and 15.2) such that long-chain alkenes tend to be liquids at room temperature whereas the corresponding alkanes are solids (compare octadecane with 1-octadecene). This effect is important in determining the states of edible lipids (see below).

The Naming of Alkenes

As we can see in Table 15.2, alkene compounds are named by substituting the suffix -ene for -ane in the name of the corresponding alkane, e.g.

$CH_3—CH_3$	$CH_2=CH_2$
ethane	ethene
$CH_3—CH_2—CH_3$	$CH_2=CH—CH_3$
propane	propene

However, more than one isomer exist for higher members of the series since the double bond may be in different positions, e.g.

$CH_3—CH_2—CH_2—CH_3$ (n–butane)

(a) $CH_3=CH—CH_2—CH_3$

(b) $CH_3—CH=CH—CH_3$

(c) $CH_3—CH_2—CH=CH_2$

butenes,

Here, (a) is identical to (c) but (b) is a different isomer. (a) is, therefore, called *1-butene* and (b) *2-butene*—the number indicating the position of the double bond. Groups attached to a main chain can be indicated as in the naming of alkanes. Thus.

$$CH_2=CH—CH_2—\overset{\displaystyle CH_3}{\overset{|}{C}H}—CH_2—CH_3$$ is called 4-methyl-1-hexene.

Properties and Reactions of Alkenes

Alkenes are very similar to alkanes in their hydrophobic properties and high heats of combustion ; however, the presence of double bonds makes alkenes much more reactive.

1. *Saturation Reactions*

The most characteristic reactions of alkenes are *saturation* reactions which involve the *addition* of functional groups to double-bonded carbon atoms, thus removing the double bond. In biochemistry and nutrition, there are two important types of saturation. Hydrogenation (addition of hydrogen) and Halogenation (addition of halogens, especially bromine and iodine).

(a) Hydrogenation, e.g.

$$CH_3—CH = CH—CH_3 + H_2 \rightarrow CH_3—CH_2—CH_2—CH_3$$

Hydrogenation reactions of alkenes have important applications in the margarine industry. The raw materials for the manufacture of margarine are oils (*liquid* lipids, e.g. groundnut, sunflower or cotton-seed oil) which are made up of large molecules containing long, unsaturated hydrocarbon chains. These oils can be 'hardened' (i.e. converted into more useful solid fats) by the saturation of the hydrocarbon chains with hydrogen, normally with the help of a finely divided platinum catalyst.

Unsaturated fatty acids are an essential part of the diet of both humans and livestock, e.g.

$$CH_3—CH_2—CH_2—CH_2—CH_2—CH = CH—CH—CH_2—CH = ——$$
$$—CH—CH_2—CH_2—CH_2—CH_2—CH_2—CH_2—CH_2—CO_2H$$

Linoleicacid

(b) Halogenation, e.g.

$$CH_3—CH = CH—CH_3 + Br_2 \rightarrow CH_3—\underset{|}{\overset{Br}{CH}}—\overset{Br}{\underset{|}{CH}}—CH_3.$$

This reaction is the basis of the standard qualitative test for the presence of multiple bonds (unsaturation) in organic compounds. Bromine dissolved in carbon tetrachloride is added dropwise to the substance under test. An unsaturated compound will react with the added bromine resulting in the disappearance of the characteristic red-brown colour of bromine.

It is essential for the food industry to have a quantitative estimate of the amount of unsaturation in lipid molecules ; this is normally expressed as the *Iodine Number*—the mass of iodine absorbed (by addition reactions) per 100 g of sample of fat or oil.

2. *Oxidation Reactions*

Under certain conditions, carbon to carbon double bound in large molecules can react with atmospheric oxygen to give a complex mixture of products. In unsaturated oils, the products have unpleasant tastes and smells. A fatty material oxidized in this way is said to be rancid and may be unfit for human consumption.

3. *Polymerization Reactions*

Two important plastics, used widely in everyday life and agriculture, are manufactured from simple alkene molecules. *Polythene* (or polyethylene) is made by the linking of many ethene molecules into long, branched chain (polymerization), i.e.

$CH_{CH_2} = CH_2 \quad CH_2 = CH_2 \quad CH_2 = CH_2$ (monomers)

$—CH_2—CH_2—CH_2—CH_2—CH_2—CH_2—$, etc. (polymer).

Polythene is the familiar plastic used for pipes, buckets, plastic bags, etc.

Polypropylene is made in a similar fashion but starting with propene as the monomer. Polypropylene is stronger than polythene and has a higher melting point. Woven polypropylene is a strong and durable substitute for hessian (jute).

Some important natural compounds containing alkene chains

(1) *Ethene* is a powerful plant growth regulator, sometimes used commercially to promote ripening in fruit.

(2) *Fatty acids.*

(3) *Many highly coloured compounds* occurring in fruit and vegetables are alkenes, e.g. Lycopene, the red colouring matter in tomatoes—

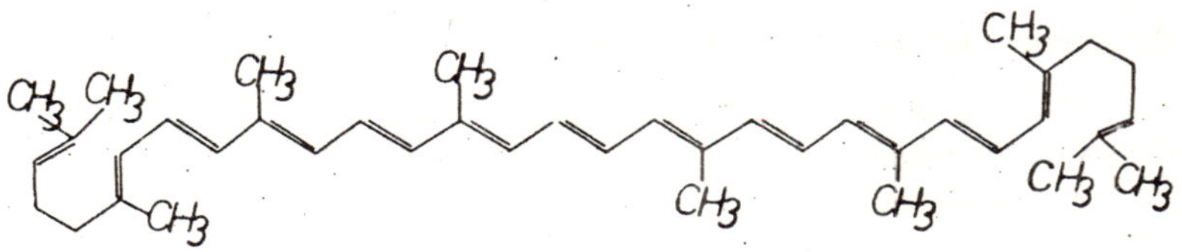

(where the C and H atoms on the main chain are omitted, for clarity).

(4) *Vitamin A1 or Retinol* (deficiency in mammals results in retarded growth and night blindness

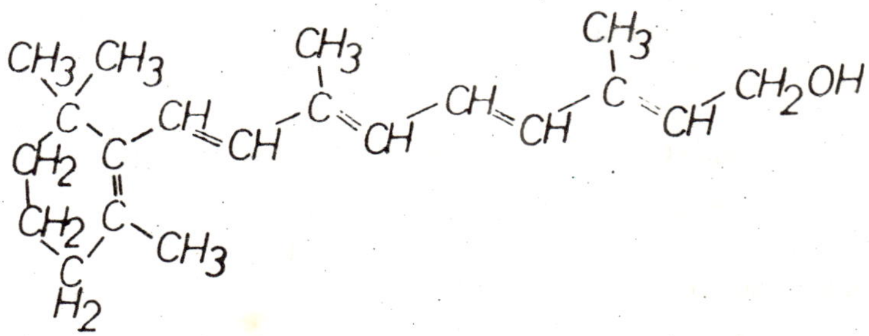

Vitamin A1

THE ALKYNES

The alkynes are straight- and branched-chain hydrocarbon compounds containing one or more triple bonds. Their general formula (1 triple bond in the molecule) is C_nH_{2n-2}, e.g.

Ethyne (Acetylene) $CH{\equiv}CH$

Propyne $CH{\equiv}C{-}CH_3$.

The Naming of Alkynes

Alkyne compounds are named by substituting the suffix -yne for -ane in the name of the corresponding alkane, e.g.

$CH_3{-}CH_3$ ethane $CH{\equiv}CH$ ethyne.

The position of the triple bond in higher members of the series is indicated in the same way as in alkenes, i.e.

$CH_3{-}CH_2{-}CH_2{-}CH_3$ butane

$CH{\equiv}C{-}CH_2{-}CH_3$ 1–butyne

$CH_3{-}C{\equiv}C{-}CH_3$ 2–butyne.

Properties and Reactions of Alkynes

The presence of triple bonds in alkynes makes them very reactive. For example, ethyne (acetylene) is an unstable, explosive gas.

1. *Combustion*

As with the other hydrocarbons, alkynes release large amounts of energy on burning. Thus in the oxyacetylene burner (for cutting and welding metals) the flow of oxygen and acetylene gases can be regulated to give flame temperatures (up to 3000°C) above the melting points of most metals.

2. *Saturation Reactions*

The most characteristic reactions of alkynes, as with alkenes, are addition reactions, resulting in the removal of triple bonds, e.g.

Hydrogenation $CH \equiv CH \xrightarrow{H_2} CH_2 = CH_2 \xrightarrow{H_2} CH_3{-}CH_3$

(normally requires a finely divided metal catalyst)

Halogenation $CH \equiv CH \xrightarrow{Br_2} CHBr = CHBr \xrightarrow{Br_2} CHBr_2{-}CHBr_2$.

Under acidic conditions and in the presence of a catalyst ($HgSO_4$), a water molecule can also be added across the triple bound :

$$CH_3{-}C{\equiv}CH + H_2O \rightarrow CH_3{-}\underset{}{\overset{OH}{\overset{|}{C}}}{-}\overset{H}{\overset{|}{C}}H.$$

The product contains the *enol group* (a hydroxyl functional group attached to a double-bonded carbon atom). This enol product is unstable and immediately undergoes a rearrangement to give a keto group :

$$CH_3{-}\overset{OH}{\overset{|}{C}}{-}\overset{H}{\overset{|}{C}}H \rightarrow \underset{\text{propanone(acetone)}}{CH_3{-}\overset{O}{\overset{\|}{C}}{-}CH_3} \qquad (\text{where}{-}\overset{O}{\overset{\|}{C}}{-}\text{is a keto group})$$

by the migration of a hydrogen atom from the hydroxyl group to the next carbon atom.

In general, the reversible reaction :

$$\text{Enol form} \rightleftharpoons \text{Keto form}$$

is called Tautomerism and the two forms are different *Tautomers* of the same compound. In the present example, almost all of the substance exists as the keto tautomer (the equilibrium position is far to the right). However, the enol tautomer may be the more stable form in other compounds, e.g. acetylacetone

$$\underset{\text{Enol}}{CH_3{-}\overset{OH}{\overset{|}{C}}{=}CH{-}\overset{O}{\overset{\|}{C}}{-}CH_3} \rightleftharpoons \underset{\text{Keto}}{CH_3{-}\overset{O}{\overset{\|}{C}}{-}CH_2{-}\overset{O}{\overset{\|}{C}}{-}CH_3}.$$

If acetylacetone existed as a simple chain molecule, it would take the keto form as in acetone. However, *in the enol form only*, a more stable ring structure can be formed by intramolecular hydrogen bonding and, therefore, 80% of the molecules of acetylacetone exist in the *enol* form—

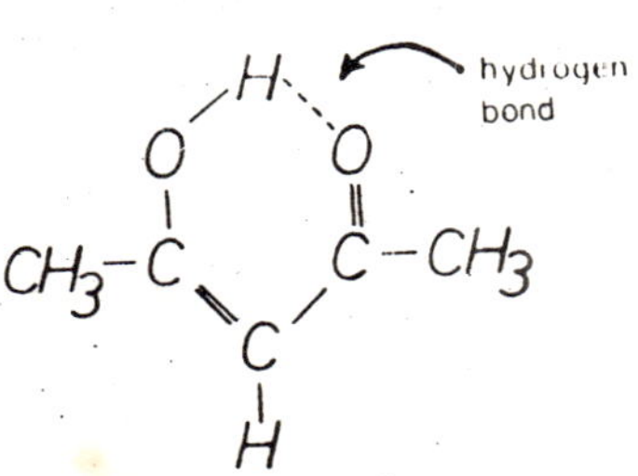

Enol $\rightleftharpoons$ keto tautomerism is very important in biochemistry.

Important Alkynes

Acetylene is by far the most common and most important alkyne, but there are a few naturally occurring, complex examples, e.g.

$$CH_3—C \equiv C—C \equiv C—C \equiv C—C \equiv C—C \equiv C—C \equiv C—CH \equiv CH_2$$

from flowers of certain species of the Compositae (Daisy family).

THE ARENES (AROMATIC HYDROCARBONS)

The simplest and most important aromatic hydrocarbon is benzene, C_6H_6. In benzene, the carbon atoms are arranged in a ring, and since carbon has a valency of 4, we might expect the carbon atoms to be linked by alternate single and double bonds (I). For simplicity, we normally draw this structure as shown (II).

I

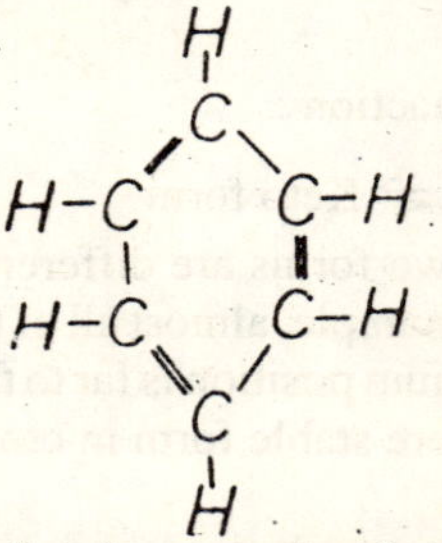

H
C
H–C C–H
H–C C–H
C
H

II

This suggests that benzene is like an alkene, and, therefore, we would expect it to behave as if it contained three normal double bonds. However, benzene behaves unusually in several ways, for example :

(1) Alkenes readily undergo *addition reactions* resulting in a saturated molecule. Benzene does not ; instead, it tends to undergo reactions in which hydrogen atoms are *replaced* by other atoms or group (*substitution reactions*). Substitution reactions normally proceed only under vigorous conditions (high temperature, high concentration of reagents, presence of a catalyst), e.g. reaction with bromine (test for unsaturation in organic molecules)—

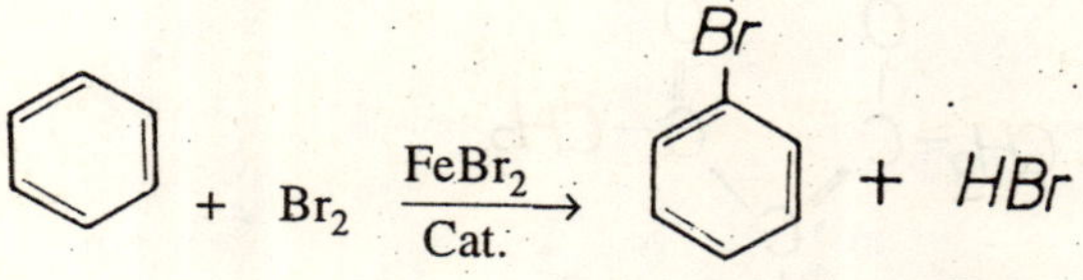

$$C_6H_6 + Br_2 \xrightarrow[\text{Cat.}]{FeBr_2} C_6H_5Br + HBr$$

NOT

$$C_6H_6 + Br_2 \longrightarrow$$ Br, H, HBr

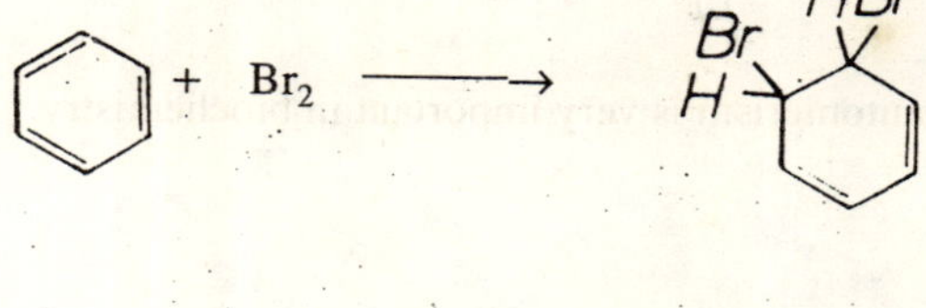

This behaviour suggests that the double bonds in benzene are not the same as those in alkenes. It also seems that the ring tends to resist saturation.

(2) When an alkene molecule reacts with hydrogen to give an alkane, e.g.

$$CH_2 = CH_2 + H_2 \rightarrow CH_3—CH_3,$$

the reaction is exergonic and results in a *more stable* (lower energy) product. Under certain conditions, benzene can also react with hydrogen to give cyclohexane :

$$C_6H_6 + 3H_2 \rightarrow C_6H_{12}.$$

However, the heat of reaction is much less than would be predicted for the saturation of three double bonds, again stressing the chemical stability (i.e. lower energy) of the benzene right bonds as compared with those in alkenes.

(3) There should be two isomers of 1, 2-dimethyl benzene—

i.e. one with a single bond between the carbon atoms carrying methyl groups and one with a double bond. In fact, these two isomers turn out to be the same compound.

The explanation of these apparent anomalies is that the bonds between carbon atoms in the benzene right are *all the same*, each being intermediate between a single and a double bond so that the valency of carbon is fulfilled. This is shown by the fact that a typical single bond between C atoms is 1·54 Å whereas the C—C bond length in benzene is intermediate, 1·39 Å (1 Å – 10^{-10} m). We can think of the true structure of benzene as being intermediate between the two structure of benzene as being intermediate between the two structures—

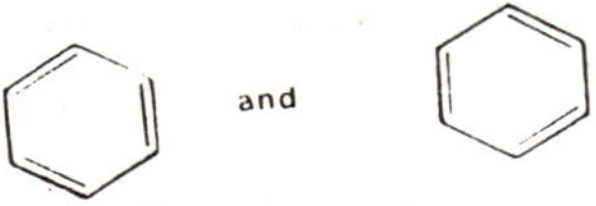

Sometimes the intermediate structure is drawn as shown (III), but we shall continue to use the conventional structure (IV)—

III or IV

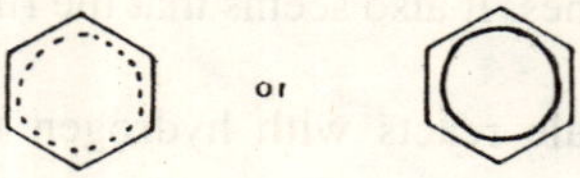

The presence of six intermediate bonds in its ring makes benzene more like an alkane than an alkene. As long as it is intact, the benzene ring is stable and unreactive.

The Naming of Substituted Benzene Compounds

Substituted benzene molecules are named by numbering the carbon atoms clockwise round the ring and using these numbers to indicate the position of attached functional groups

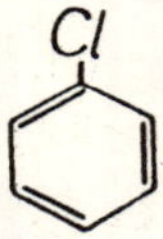

chlorobenzene

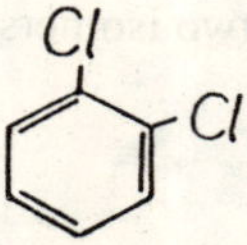

1,2-dichlorobenzene (or ortho-dichlorobenzene)

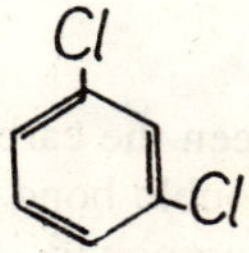

1,3-dichlorobenzene (or meta-dichlorobenzene)

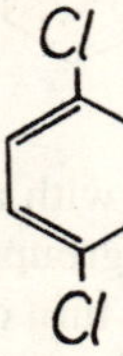

1,4-dichlorobenzene (or para-dichlorobenzene)

Chemical Properties and Reactions of Arenes

Aromatic hydrocarbons derive their name (aromatic) from their strong, usually pleasant smells. Most are toxic and several, including benzene, are carcinogenic (cause cancer). Arenes are highly hydrophobic, and benzene is a very effective (but dangerous) solvent for grease. In general, aromatic hydrocarbons are chemically inert, although, as well shall see later, the presence of attached atoms or groups can make them more reactive.

The Importance and Occurrence of Arenes

Unsubstituted arene molecules do not normally occur in living organisms ; in fact, we have noted above that these compounds are toxic to humans. On the other hand, there are countless substituted aromatic compounds involved in biochemical pathways ; for example, three amino acids necessary for the synthesis of proteins contain benzene rings, although the mammalian body cannot synthesize the benzene ring. Consequently, these three amino acids, Phenylalanine, Tryptophan and Tyrosine are amongst the *essential amino acids*, which must be derived (directly or indirectly) from plants. The unreactivity of aromatic rings also has important consequences in several branches of environmental science ; in particular it contributes to the persistence in the soil (resistance to chemical attack) of some aromatic pesticides and to the stability of soil humus.

Pure aromatic hydrocarbons can be synthesized industrially or obtained from petroleum or coal. Various substituted arenes are important in the industrial production of dyes, explosives, plastics, drugs, disinfectants, etc.

Arenes other than Benzene

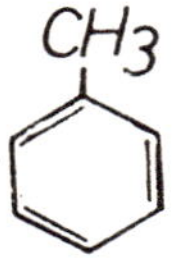

CH_3

Toluene

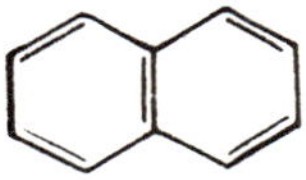

Naphthalene

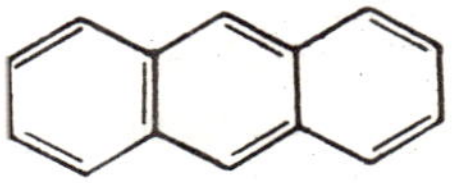

Anthracene

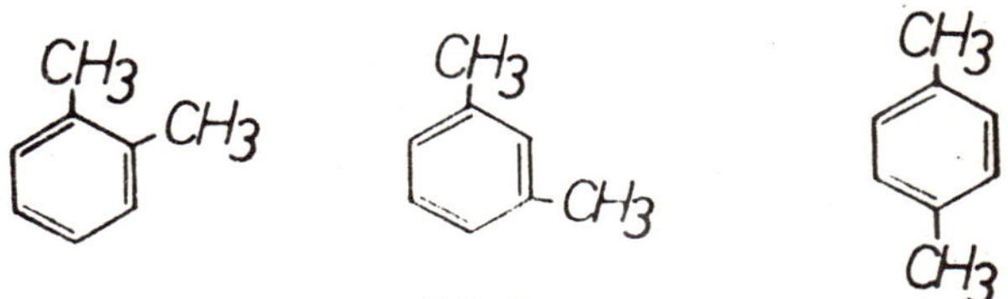

CH_3 CH_3

CH_3 CH_3

CH_3 CH_3

Xylenes

Note that there are aromatic compounds which include atoms other than carbon in the ring. these compounds are called *Heterocyclic e.g.*—.

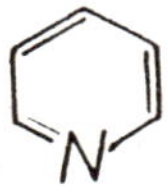

N

Pyridine

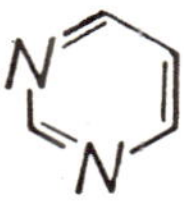

N N

Pyrimidine

CHAPTER 16

Organic Compounds Containing Oxygen

In Chapter 15, we considered the four main groups of hydrocarbons, which contain carbon and hydrogen atoms only, but most biochemical substances (carbohydrates, lipids etc.) also have oxygen-containing functional groups attached to the hydrocarbon frame-work. In this chapter, we shall look at the properties and reactions associated with these function groups.

THE ALCOHOLS

The alcohols are aliphatic (*not* aromatic) compounds containing one or more hydroxyl groups attached to a hydrocarbon chain, which may be straight or branched, saturated or unsaturated.

The Naming of Alcohols

Using the systematic method of naming organic molecules, we delete the final -e in the name of the corresponding alkane and substitute -ol. Thus,

CH_4 (Methane) gives CH_3OH (Methanol)

and

$$CH_3-CH_2-\underset{|}{\overset{CH_3}{CH}}-CH_2-CH_3 \quad \text{gives} \quad CH_3-CH_2-\overset{CH_3}{\underset{|}{CH}}-CH_2-CH_2-OH$$

3–methyl–pentane gives 3 – methyl–l–pentanol

where the number preceding the basic name (pentanol) indicates the position of the hydroxyl group. We shall not normally require to name more complicated alcohols (highly branched and unsaturated).

The carbon atom to which the hydroxyl group is attached may be called the α-carbon atom and it is possible to classify alcohols according to the number alkyl groups (written as R groups) attached to the α-carbon, i.e.

Primary alcohols (1 alkyl group)

$$R-\overset{\displaystyle H}{\underset{\displaystyle H}{\overset{|}{\underset{|}{C}}}}-OH$$

e.g. CH_3-CH_2-OH

Secondary alcohols (2 alkyl groups)

$$R-\overset{\displaystyle R}{\underset{\displaystyle H}{\overset{|}{\underset{|}{C}}}}-OH$$

$$\text{e.g. } CH_3-\overset{\displaystyle CH_3}{\overset{|}{C}H}-OH$$

Tertiary alcohols (3 alkyl groups)

$$R-\overset{\displaystyle R}{\underset{\displaystyle R}{\overset{|}{\underset{|}{C}}}}-OH$$

$$\text{e.g. } CH_3-\overset{\displaystyle CH_3}{\underset{\displaystyle CH_3}{\overset{|}{\underset{|}{C}}}}-OH$$

Table 16.1. Properties of some straight-chain alcohols.

Name	*Formula*	*Molecular weight*	*m.p. °C*	*b.p. °C*	*Water solubility (g 100 mol^{-1})*
Methanol	$CH_3\,OH$	32	-94	65	*
Ethanol	$C_2\,H_5\,OH$	46	-117	79	*
1-propanol	$C_2\,H_7\,OH$	60	-127	97	*
1-butanol	$C_4\,H_9\,OH$	74	-90	117	8.0
1-pentanol	$C_5\,H_{11}\,OH$	88	-79	137	2.2
1-hexanol	$C_6\,H_{13}\,OH$	102	-47	158	0.6
1-heptanol	$C_7\,H_{15}\,OH$	116	-34	176	0.1
1-octanol	$C_8\,H_{17}\,OH$	130	-17	195	0.04
1-decanol	$C_{10}\,H_{21}\,OH$	158	7	229	0.004

*Soluble in all proportions of water to alcohol.

Properties and Reactions of Alcohols

1. *Physical properties*

If we compare the melting and boiling points of alcohols (Table 16.1) with those of the corresponding alkanes, we find that the alcohols have much higher values, although the difference diminish as the molecules become longer. One result of these higher values is that there are no gaseous alcohols.

The relatively high temperature for change of state and the very higher water solubilities of the lower alcohols (C1 to C5) are due to the formation of strong hydrogen bonds between the hydroxyl groups of alcohol molecules and also between hydroxyl groups and water molecules, e.g.

$$CH_3-O-H\text{-----}\underset{\underset{CH_3}{|}}{\overset{\overset{H}{|}}{O}}\qquad CH_3-O-H\text{-----}\underset{\underset{H}{|}}{\overset{\overset{H}{|}}{O}},\text{ etc.}$$

However, because alcohol molecules are amphipathic (contain a hydrophilic hydroxyl group and a hydrophobic hydrocarbon chain), as the hydrocarbon chain becomes longer, its hydrophobic properties come to dominate the properties of the molecule. Consequently, the higher alcohols (C_{10} upwards) tend to resemble closely the corresponding alkanes (unreactive and insoluble viscous liquids and waxy solids).

2. *Acid-base Properties.*

Alcohols are *very* weak acids. For the ionization

$$ROH + H_2O \rightleftharpoons RO^- + H_3O^+$$

K_a values vary from 10^{-13} to 10^{-18} depending upon the length and nature of R. Under certain vigorous conditions, salts of Alcohols can be formed, e.g. the reaction of pure alcohols with alkali metals :

$$2ROH + 2Na \rightarrow 2Na^+RO^- + H_2\uparrow$$

3. *Redox Reactions*

These reactions are used in the laboratory to distinguish between primary, secondary and tertiary alcohols.

Primary Alcohols are oxidized by strong oxidizing agents (e.g. potassium dichromate or potassium permanganate) to give, first, aldehydes and, subsequently, carboxylic acids, e.g.

$$CH_3CH_2OH \rightarrow CH_3\overset{\overset{\displaystyle O}{\|}}{C}—H \rightarrow CH_3\overset{\overset{\displaystyle O}{\|}}{C}—OH$$

Ethanol Ethanol (Ethanoic acid)
(Acetaldehyde) (Acetic Acid.)

A practical example of this reaction is the production of sour wine or beer (containing acetic acid) when a fermentation mixture becomes contaminated with ethanol-oxidizing micro-organisms.

Secondary Alcohols can be oxidized to ketones, but no further (because formation of the acid would involve breaking strong C—C bonds), e.g.

$$CH_3\ \overset{\overset{\displaystyle OH}{|}}{C}H\,CH_3 \rightarrow CH_3\ \overset{\overset{\displaystyle O}{\|}}{C}\ CH_3 \text{ Acid.}$$

2–propanol Propanone

Teritary alcohols cannot be oxidized by these reagents.

4. *Esterification*

Alcohols react with carboxylic acids to give *esters*. The general reaction may be written :

$$R—OH + R^1—\overset{\overset{\displaystyle O}{\|}}{C}—OH \rightarrow R^1—\overset{\overset{\displaystyle O}{\|}}{C}—OR + H_2O.$$

For example, the reaction of ethanol with ethanoic (acetic) acid

$$C_2H_5—OH + CH_3—\overset{\overset{\displaystyle O}{\|}}{C}—OH \rightarrow CH_3—\overset{\overset{\displaystyle O}{\|}}{C}—OC_2H_5 + H_2O$$

yields the fruit-smelling ester ethyl acetate which is a versatile solvent for organic substances (e.g. in glues and varnishes, and in dry cleaning textiles). Easters play a part in many biochemical path-ways, notably in energy and lipid metabolism, as well as acting, due to their strong pleasant smells, as chemical attractant in nature (i.e. to ripe fruit, and between insects).

5. *Formation of Hemiacetals and Condensation of Alcohols*

Two further types of reaction are important in the structure and chemistry of carbohydrates (sugars, starch, cellulose, etc.) which are polyhydroxy alcohols containing aldehyde and ketone groups. Firstly, an alcohol can react with an aldehyde to give a hemiacetal :

$$R-\overset{\overset{\large O}{\|}}{C}-H + R^1OH \rightarrow R-\underset{\underset{\large H}{|}}{\overset{\overset{\large OH}{|}}{C}}-OR^1$$

Aldehyde Alcohol Hemiacetal

A similar reaction occurs with ketones. Where both aldehyde and alcohol functional groups are on the same molecule, this (intramolecular) reaction gives a ring structure.

As a result of the second reaction type, the condensation, or polymerization, of two (or more) alcohol molecules,

$$R-OH + HO-R^1 \rightarrow R-O-R^1 + H_2O$$

many single monosaccharide units can be joined together to give long polysaccharide chains.

Ethanol (Ethyl Alcohol)

Ethanol is very useful in the laboratory as a dehydrating solvent, as a dehydrating agent in preparing material for microscopy and as a preservative for plant and animal specimens. It is also the important ingredient in alcoholic drinks—beer (about 4%), wine (10-20%) and spirits (about 40%).

Three main types of ethanol are used in the laboratory :

Absolute — 100% ethanol ;

Rectified — 95% ethanol (the remaining 5% including water and benzene) ;

Denatured — as for rectified but containing methanol, 2-propanol or benzene to make it unfit for human consumption e.g. methylated spirits contains methanol.

The concentration of ethanol in alcoholic drinks is normally expressed as *degrees proof* (approximately twice the percentage composition).

Ethanol is a common end-product of anaerobic respiration (fermentation) in higher plants (as in other organism) ; its effects on cell membranes are though to be at least partly responsible for the damaging effect of flooding on sensitive plant species.

Other Important Alcohols

Methanol, CH_3OH (wood alcohol), may be obtained by heating certain hardwoods and condensing the vapour produced. It is highly toxic, causing blindness in humans. Methanol is used as a solvent, for denaturing ethanol and in the synthesis of formaldehyde.

Ethylene glycol, $\underset{OH}{\underset{|}{CH_2}}-\underset{OH}{\underset{|}{CH_2}}$, is dihydroxy alcohol used as a solvent, as anti-freeze in vehicle radiators and as a coolant liquid for refrigerators.

Glycerol, $\underset{OH}{\underset{|}{CH_2}}-\underset{OH}{\underset{|}{CH}}-\underset{OH}{\underset{|}{CH_2}}$ (glycerine), is a sweet-tasting, viscous trihydroxy alcohol. It is an important constituent of lipids and is a valuable by-product of soap manufacture. Due to its hygroscopic nature, glycerol is mixed with tobacco to keep it moist.

Monosaccharides are polyhydroxy alcohols, e.g. glucose

$$H-\overset{O}{\overset{\|}{C}}-\underset{H}{\underset{|}{\overset{OH}{\overset{|}{C}}}}-\underset{OH}{\underset{|}{\overset{H}{\overset{|}{C}}}}-\underset{H}{\underset{|}{\overset{OH}{\overset{|}{C}}}}-\underset{H}{\underset{|}{\overset{OH}{\overset{|}{C}}}}-\underset{H}{\underset{|}{\overset{OH}{\overset{|}{C}}}}-H$$

Long-chain alcohols are used as 'detergents.

Many other biochemical molecules contain the hydroxyl groups, e.g. Vitamin Al.

THE PHENOLS

The phenols are similar to alcohols except that the hydroxyl functional groups is attached directly to an aromatic ring. The simplest member of the series is phenol (carbolic acid) (I)—

I C_6H_5OH II $C_6H_4(OH)_2$

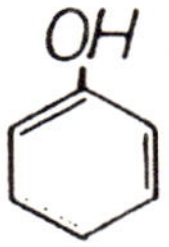

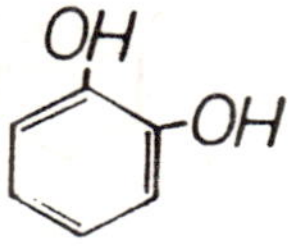

Aromatic compounds are commonly given special, unsystematic names. One example (II), is normally called Catechol, although correctly it is 1, 2-dihydroxy benzene or orthodihydroxy benzene.

Properties and Reactions of Phenols

Although phenols behave like aliphatic alcohols in a few reaction(e.g. esterification with carboxylic acids), the influence of the aromatic ring makes the chemistry of the hydroxyl groups generally very different in phenols.

1. *Acid/base Properties*

Because the benzene ring tends to *attract electrons*, the hydrogen atom in a phenol hydroxyl groups is not held as tightly as the hydrogen atom in an aliphatic alcohol hydroxyl group–

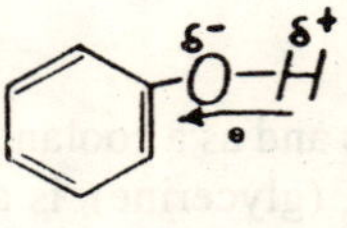

Consequently the equilibrium position in the reaction—

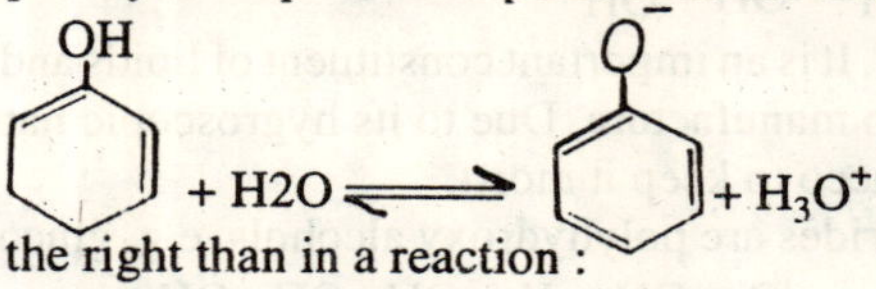

$$C_6H_5OH + H2O \rightleftharpoons C_6H_5O^- + H_3O^+$$

is further to the right than in a reaction :

$$R—OH + H_2O \rightarrow RO^- + H_3O^+$$

with the result that phenols are more acidic than alcohols, as shown by the following K_a values :

Phenol	10^{-10}
Aliphatic alcohols	10^{-13} to 10^{-18}

Phenol is normally considered to be an acid and readily forms salts with alkalis and alkali metals. Alcohols are not acids.

2. *Redox Reactions*

Phenols are more readily oxidized than alcohols e.g.—

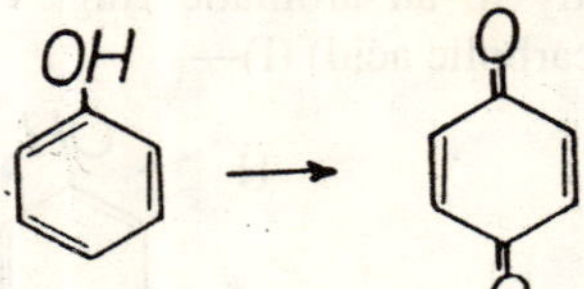

$$C_6H_5OH \longrightarrow \text{p-benzoquinone}$$

The products of oxidation are normally highly coloured *quinones* (cyclic diketones). This type of redox reaction, which does not require strong oxidizing agents, is employed in photographic developing, where the silver bromide on an exposed film is precipitated as silver by the redox reaction—

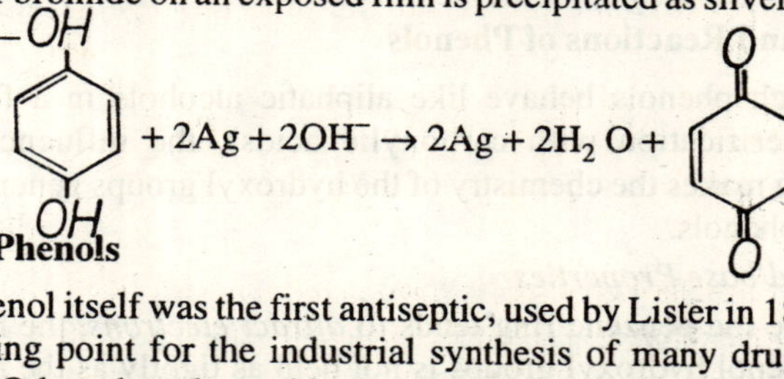

$$HO-C_6H_4-OH + 2Ag + 2OH^- \rightarrow 2Ag + 2H_2O + O{=}C_6H_4{=}O$$

Uses of Phenols

Phenol itself was the first antiseptic, used by Lister in 1865. It is also the starting point for the industrial synthesis of many drugs, dyes and plastics. Other phenols used in the chemical industry include—

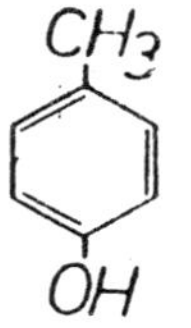

p-cresol 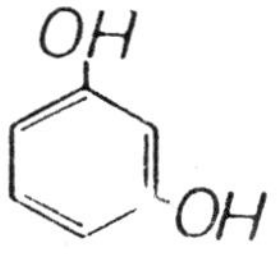 Resorcinol 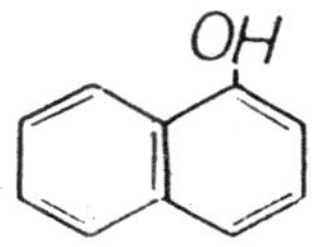 1-naphthol (α-naphthol)

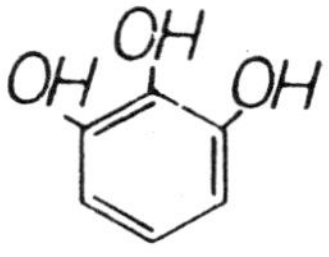

Pyrogallol 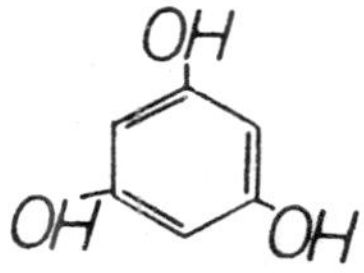 Phloroglucinol

THE ETHERS

In an ether molecule, two aliphatic groups, two aromatic groups (Ar), or one of each type are linked together through a single oxygen atom.

i.e. R—O—R, Ar—O—Ar R—O—Ar.

e.g. CH_3—O—C_2H_5,

methyl ethyl ether

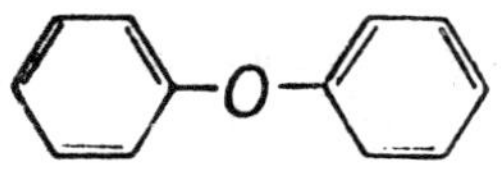

Diphenyl ether 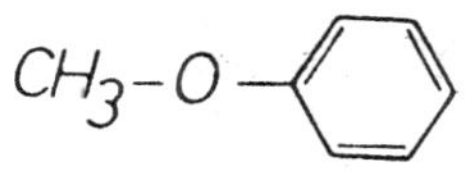 Methyl phenyl ether (Anisole)

As indicated, ethers are named simply by listing the two groups attached to the oxygen atom followed by the word ether. (Note the name *phenyl* for an attached benzene ring).

Properties and Reaction of Ethers

The outstanding property of ethers is their unreactivity and, because of this, they are commonly used as inert solvents for organic reactions. However, ethers can react explosively in the presence of light and oxygen and for this reason they are normally kept in brown bottles to exclude light.

The Occurrence and Uses of Ethers

The ether link occurs in many biochemical molecules, for example in, and between, the rings of carbohydrates and in natural products such as—

Vanillin
(vanilla)

Diethyl ether, $C_2H_5OC_2H_5$, sometimes simple called ether, is probably the most important of the series because of its use as a solvent and anaesthetic. However, diethyl ether must be handled with the greatest care ; the high density of its vapour (greater than air) can result in the accumulation of pockets of inflammable gas, for example, in laboratory drains.

Other ethers, important as solvents and in the synthesis of organic compounds, include—

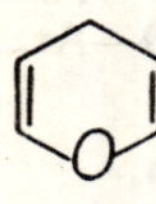
α-pyran

Furan

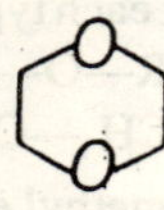
Dioxane

Note that the names *pyranose* and furanose for 6- and 5-membered monosaccharide rings originate from pyran and furan.

ALDEHYDES AND KETONES

Aldehyde and ketone molecules contain the carbonyl functional group,

$$\overset{\displaystyle O}{\overset{\|}{—C—}}$$

In aldehydes, the group is at the end of a chain, e.g. Ethanal (acetaldehyde)

$$CH_3—\overset{\displaystyle O}{\overset{\|}{C}}—H$$

whereas in ketones, it occurs within the chain, e.g. Propanone (acetone, dimethyl ketone)

$$CH_3—\overset{\displaystyle O}{\overset{\|}{C}}—CH_3$$

We shall see that carboxylic acids and acid derivatives also carry the carbonyl group, but in close association with another functional group.

Naming Aldehydes and Ketones

Aldehydes—using the systematic method, we delete the final-e in the name of the corresponding alkane and substitute-al.

Thus $\underset{\text{Ethane}}{C_2H_6}$ gives $\underset{\text{Ethanal}}{CH_3CHO}$

and

$$\underset{\text{3–methyl–pentane}}{CH_3CH_2—\overset{\overset{\large CH_3}{|}}{CH}—CH_2—CH_3}$$

gives

$$\underset{\text{3–methyl–pentanal}}{CH_3CH_2—\overset{\overset{\large CH_3}{|}}{CH}—CH_2—CHO}$$

where the aldehyde-group-carbon atom is always numbered Cl.

Ketones—using the systematic method we delete the final -e of the name of the corresponding alkane and substitute -one, e.g.

$$\underset{\text{Butane}}{CH_3CH_2CH_2CH_3} \quad \text{gives} \quad \underset{\text{2–butanone}}{CH_3\overset{\overset{\large O}{||}}{C}CH_2CH_3}$$

where the position of the carbonyl group is indicated by the number preceding the main name e.g.

$$\underset{\text{3–chloro–2–pentanone.}}{CH_3CH_2—\overset{\overset{\large Cl}{|}}{CH}—\overset{\overset{\large O}{||}}{C}CH_3}$$

Ketones may also be named like ethers, although this is a more clumsy method, e.g

$$CH_3—\overset{\overset{\large O}{||}}{C}—CH_2 \qquad \text{Dimethyl Ketone}$$

Table 16.2. Properties of some aldehydes and ketones.

Name	*Formula*	*Molecular weight*	*m.p. °C*	*b.p. °C*	*Water solubility ($g\ 100\ ml^{-1}$)*
Aldehydes					
Methanal (Formaldehyde)	HCHO	30	-92	-21	Very soluble
Ethanal (Acetaldehyde)	CH_3CHO	44	-121	21	*
Propanal	C_2H_5CHO	58	-81	49	16
Butanal	C_3H_7CHO	72	-99	76	4
Pentanal	C_4H_9CHO	86	-92	103	Slightly soluble
Ketones					
Propanone (Acetone)	CH_3COCH_3	58	-95	56	*
2-butanone	$CH_3COC_2H_5$	72	-86	80	33
2-pentanone	$CH_3COC_3H_7$	86	-78	102	6
3-pentanone	$C_2H_5COC_2H_5$	86	-40	102	5
2-hexanone	$CH_3COC_4H_9$	100	-57	128	1.6

* Soluble in all proportions.

Properties and Reactions of Aldehydes and Ketones

As with most organic compounds, the physical properties of aldehydes and ketones are a compromise between the hydrophobic and dehydrating parts of the molecule (see Table 16.2). Aldehydes and ketones (like esters) tend to have pleasant smells and flavours (e.g. Vanillin which is an aldehyde as well as an ether). Some of the chemical properties of aldehydes and ketones have been described in previous sections, i.e. Keto/Enol tautomerism and the formation of Hemiacetals.

1. *Redox Reactions*

Alipathic aldehydes can be oxidized to the corresponding carboxylic acid, using either Fehling's or Tollen's reagent, whereas aliphatic ketones can not. This is an important method for discriminating between the two types of compound and is the basis for the classification of monosaccharides into reducing (aldehyde) and non-reducing (ketone) sugars, e.g.

$$\underset{\text{Ethanal}}{CH_3-\overset{\overset{O}{\|}}{C}-H} \rightarrow \underset{\text{Ethanoic acid}}{CH_3-\overset{\overset{O}{\|}}{C}-OH}$$

$$\underset{\text{Propanone}}{CH_3.-\overset{\overset{O}{\|}}{C}-CH_3}$$

2. *Condensation by the Aldol Reaction*

Under alkaline conditions, two molecules of ethanal can *condense* together :

$$2CH_3\overset{\overset{\displaystyle O}{\|}}{C}—H \rightarrow CH_3\overset{\overset{\displaystyle OH}{|}}{CH}CH_2\overset{\overset{\displaystyle O}{\|}}{C}—H$$

to give 3-hydroxy-butanal. This reaction is a simple example of an *Aldol Condensation*, and it can proceed further to give a long chain of carbon atoms. In a slightly different form, this reaction is involved in the biosynthesis of long-chain fatty (carboxylic) acid from 2-carbon molecules.

3. *Polymerization Reactions*

In aqueous solution, methanal (formaldehyde) tends to polymerize to give a long-chain compound (paraformaldehyde) which is the active ingredient of the biological preservative, formalin, i.e.

$$nH—\overset{\overset{\displaystyle O}{\|}}{C}—H + H_2O \rightarrow HO—CH_2—(OCH_2)_{n-2}O—CH_2OH$$

On heating, the gaseous monomer methanal is again released. Under other conditions, a trimer can be formed—

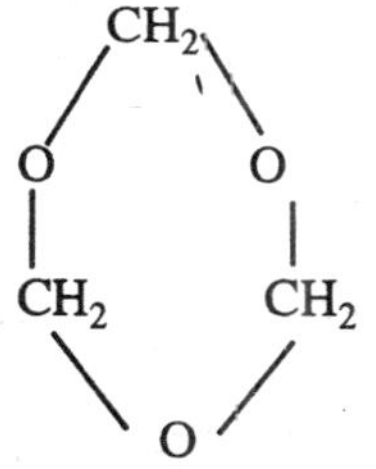

CH_2 (top), O, O, CH_2, CH_2, O (bottom) — six-membered ring

Trioxymethylene

Similar polymerization reactions occur with ethanal giving a trimer or a tetramer—

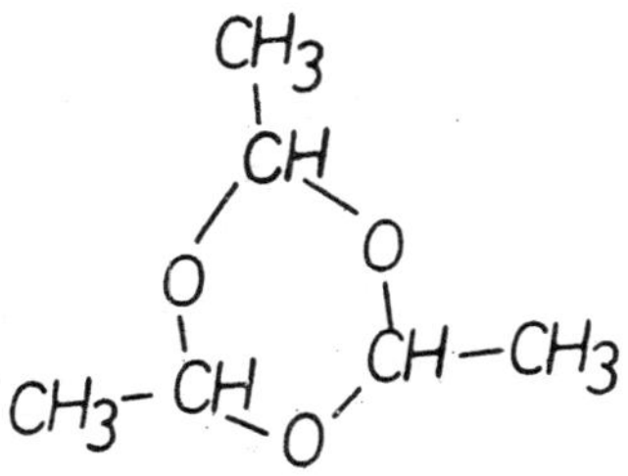

Paraldehyde: ring of three $CH(CH_3)$ groups alternating with three O atoms

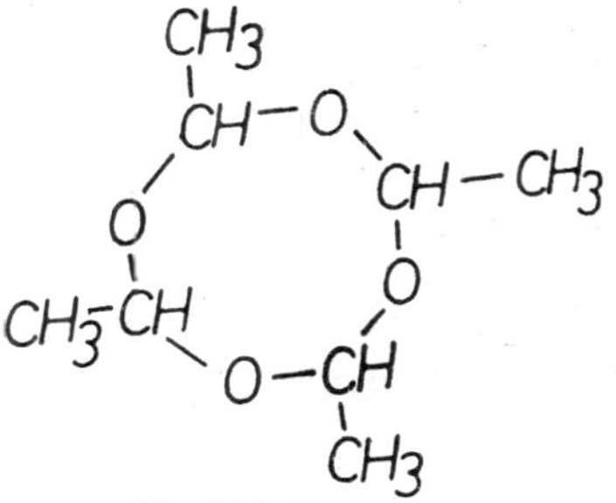

Metaldehyde: ring of four $CH(CH_3)$ groups alternating with four O atoms

Paraldehyde is a sleep-inducing drug where as metaldehyde is a slug and snail poison (molluscicide).

Important Aldehydes and Ketones

Methanal (formaldehyde), in the form of formation (37% methanal in water) is used as a disinfectant and preservative. It is also used in the tanning of hides, the synthesis of plastics and drugs, as a fumigant and in the manufacture of mirrors. Methanal is a gas at room temperature.

Propanone (acetone) is a useful solvent. It is a normal intermediate product of the breakdown of lipids in humans and is subsequently oxidized to CO_2 and H_2O. In diabetes, acetone is produced in large quantities in the body and may be smelt on the breath. High ketone concentrations also occur in the blood of cattle as a result of perinatal diseases.

Long-chain aldehydes are used in perfumes.

Monosaccharide molecules carry an aldehyde or ketone group in addition to several hydroxyl groups.

Aromatic Ketones—The Quinones

Phenols are readily oxidized to quinones, cyclic diketones

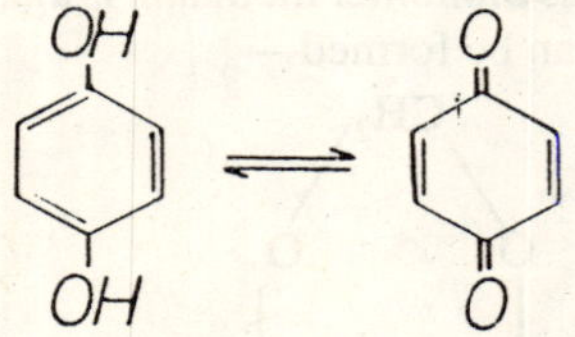

OH OH ⇌ O O

Quinones do not resemble aliphatic ketones in their properties or reactions. Their most characteristic reactions are reversible redox reactions (see above) which result in dramatic colour changes.

The strong colours of quinones are used in dyeing. For example is

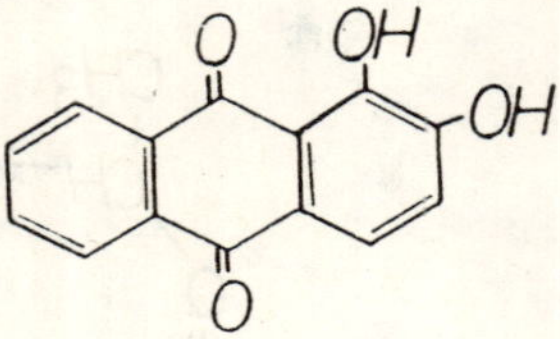

O OH OH O

the structure of Alizarin, the 'Turkey Red' dye extracted from the madder plant. The quinones structure occurs in many other natural molecules ; e.g.—

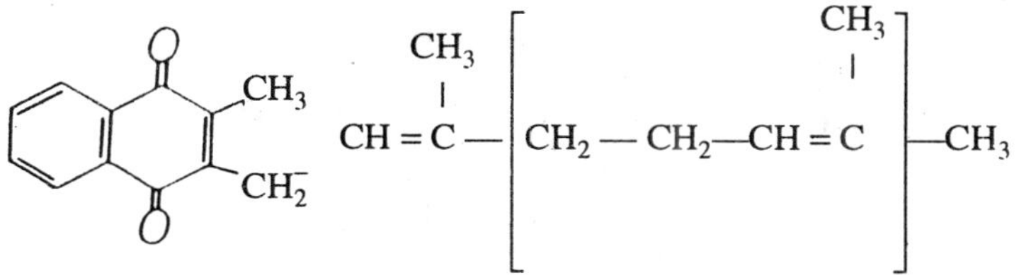

O, O, CH_3, CH_2^-

$$CH=\overset{\displaystyle CH_3}{\overset{|}{C}}-\left[CH_2-CH_2-CH=\overset{\displaystyle CH_3}{\overset{|}{C}}\right]-CH_3$$

Vitamin K1

CARBOHYDRATES

Carbohydrates can be defined formally as polyhydroxy aldehydes or ketones (or substances which can be hydrolysed to give polyhydroxy aldehydes or ketones, thus including compounds like phosphorylated sugars and amino sugars). They are particularly important in supplying energy for growth and metabolism, and in providing structural support in plants. The basic formula of all carbohydrates is $(CH_2\,O)_n$ where n can vary from 3 up to many hundreds or thousands. Whatever the size of a carbohydrate molecule, it is made up of simple units called monosaccharide molecules.

The Monosaccharides

Monosaccharide molecules normally contain from two to six carbon atoms to which are attached one carbonyl and two or more hydroxyl groups.

Example (Note the system of numbering C atoms which is the same as that for aldehydes and ketones.)

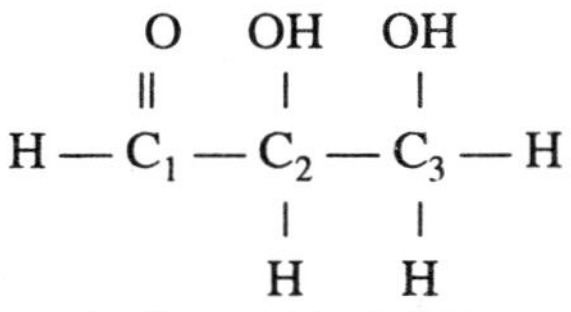

$$H-\overset{\displaystyle O}{\overset{\|}{C_1}}-\underset{\displaystyle H}{\underset{|}{\overset{\displaystyle OH}{\overset{|}{C_2}}}}-\underset{\displaystyle H}{\underset{|}{\overset{\displaystyle OH}{\overset{|}{C_3}}}}-H$$

1. Glyceraldehyde (C3)

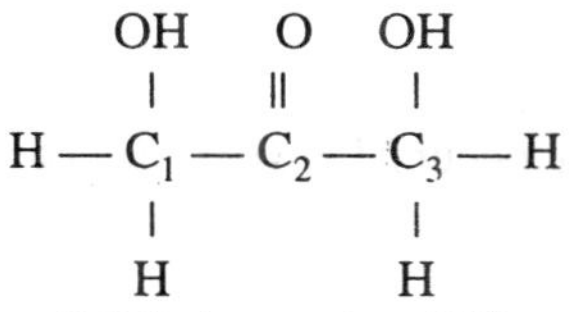

$$H-\underset{\displaystyle H}{\underset{|}{\overset{\displaystyle OH}{\overset{|}{C_1}}}}-\overset{\displaystyle O}{\overset{\|}{C_2}}-\underset{\displaystyle H}{\underset{|}{\overset{\displaystyle OH}{\overset{|}{C_3}}}}-H$$

2. Dihydroxyacetone (C3)

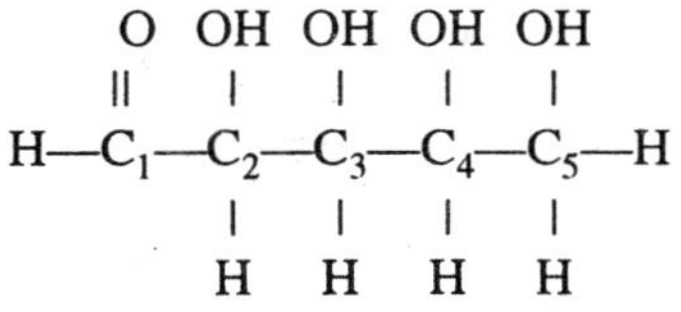

$$H-\overset{\displaystyle O}{\overset{\|}{C_1}}-\underset{\displaystyle H}{\underset{|}{\overset{\displaystyle OH}{\overset{|}{C_2}}}}-\underset{\displaystyle H}{\underset{|}{\overset{\displaystyle OH}{\overset{|}{C_3}}}}-\underset{\displaystyle H}{\underset{|}{\overset{\displaystyle OH}{\overset{|}{C_4}}}}-\underset{\displaystyle H}{\underset{|}{\overset{\displaystyle OH}{\overset{|}{C_5}}}}-H$$

3. Ribose (C5)

$$H-\underset{\displaystyle H}{\underset{|}{\overset{\displaystyle OH}{\overset{|}{C_1}}}}-\overset{\displaystyle O}{\overset{\|}{C_2}}-\underset{\displaystyle OH}{\underset{|}{\overset{\displaystyle H}{\overset{|}{C_3}}}}-\underset{\displaystyle H}{\underset{|}{\overset{\displaystyle OH}{\overset{|}{C_4}}}}-\underset{\displaystyle H}{\underset{|}{\overset{\displaystyle OH}{\overset{|}{C_5}}}}-\underset{\displaystyle H}{\underset{|}{\overset{\displaystyle OH}{\overset{|}{C_6}}}}$$

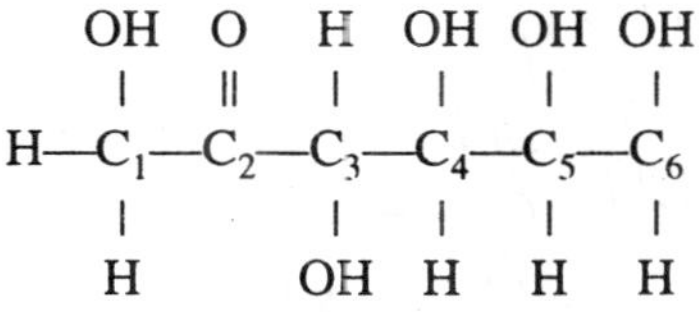

4. Fructose (6C)

$$\begin{array}{ccccccccccccc} & & O & & OH & & H & & OH & & OH & & OH \\ & & \| & & | & & | & & | & & | & & | \\ H & - & C_1 & - & C_2 & - & C_3 & - & C_4 & - & C_5 & - & C_6 & - H \\ & & & & | & & | & & | & & | & & | \\ & & & & H & & OH & & H & & H & & H \end{array}$$

5. Glucose (6C)

Molecules with 3 carbon atoms (including the carbonyl C atom, example 1 and 2) are called *trioses*, those with 5 (example 3) *pentoses* and those with 6 (example 4 and 5), *hexoses*. Monosaccharides can be further classified into *reducing sugars*, carrying a ketone group (Ketoses). Reducing sugars give positive Fehling's and Tollen's tests, whereas non-reducing do not.

Most of the properties of monosaccharides can be predicated from those of alcohols and carbonyl compounds. In particular, they are highly soluble in water due to the presence of several hydrophilic groups. Many monosaccharides and disaccharides have sweet tastes.

Ring Structures of Monosaccharides

We have drawn our examples of pentose and hexose molecules as straight chains. However, these molecules normally exist in ring form due to an intramolecular hemiacetal reaction. For example, in glucose, instead of a reaction between an aldehyde group on one molecule and an alcohol on another :

$$R - OH + R^1 - \overset{\overset{\displaystyle O}{\|}}{C} - H \rightarrow R - O - \underset{\underset{\displaystyle H}{|}}{\overset{\overset{\displaystyle OH}{|}}{C}} - R^1$$

the reaction occurs between the two ends (Cl and C5) of the same molecule—

$$\begin{array}{c} CH_2OH\ (C_6) \\ | \\ H - C_5 - OH \\ | \\ H - C_4 - OH \\ \ldots \\ C_3(OH)(H) - C_2(H)(OH) - C_1H{=}O \end{array} \longrightarrow \begin{array}{c} CH_2OH\ (C_6) \\ | \\ C_5 - O - C_1(OH)(H) \\ C_4(H)(OH) \quad C_3(OH)(H) - C_2(H)(OH) \end{array}$$

β-form

to give a six-membered pyramose ring. Glucose normally exists in the Pyranose ring from but the pyranose ring form but the ether linkage in the ring can break readily to give the chain form, when necessary.

Two important aspects of this representation of the pyranose ring should be stressed :

(a) The drawing indicates that the OH group on Cl (as well as H on C2, OH on C3, H on C4 and CH_2OH on C5) is above the plane of the ring, whereas the H atom on Cl (as well as the other groups of the molecule) is held below the plane of the ring. This is the β form of the pyranose ring. The hemiacetal reaction can also give the α form in which the hydroxyl group at C_1 is held below the ring—

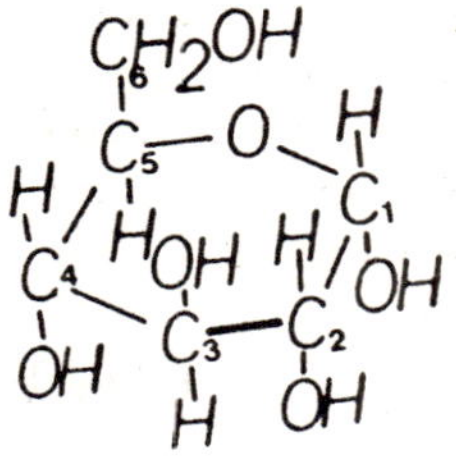

CH_2OH

α–form

(b) Although, for convenience, the six-membered ring is drawn as if it were flat, it is in face a puckered ring. However, the five-membered furanose ring form adopted by both fructose (6 C) and ribose (5C) is planar—

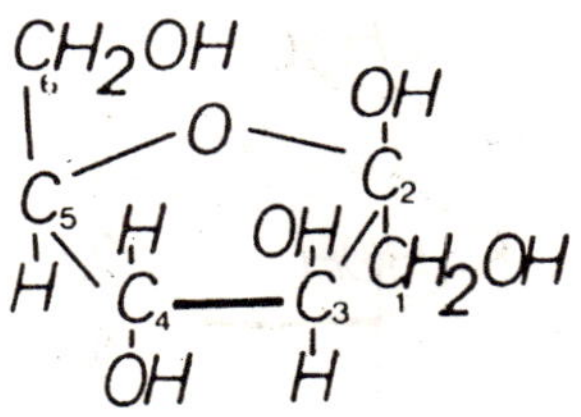

Fructose

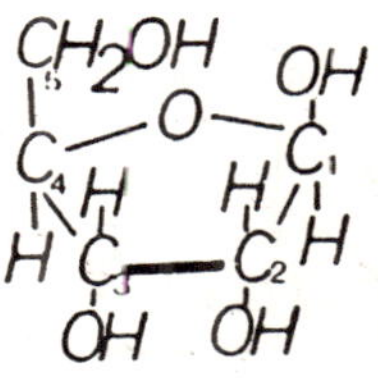

Ribose (β)

(In future diagrams, the C atoms will be omitted, for clarity.)

Although they can form pyranose rings, pentoses like ribose tend to exist in the furanose form. Amongst hexoses, glucose tends to adopt the six-membered, the fructose the five-membered ring.

Importance of Monosaccharides

Glucose is the most commonly occurring monosaccharide (in fruit, starch, cellulose, glycogen, etc.). It is enormously important in biochemistry.

Fructose occurs in fruits and, with glucose, in sucrose.

Ribose molecules are components of RNA and DNA.

The Disaccharides

The alcohol molecules can condense together as follows :

$$R—OH + HO—R^1 \rightarrow R—O—R^1 + H_2O$$

A similar reaction between hydroxyl groups on two adjacent monosaccharide molecule can lead to the synthesis of a disaccharide molecule. Since monosaccharide molecules are polyhydroxy alcohols, there are many possible ways in which the two molecules could link together. However, in nature, the link is normally formed between the Cl of one molecule and the C2, C4 or C6 on the other.

For example, two glucose molecules (α form) can condense together to give the disaccharide, maltose—

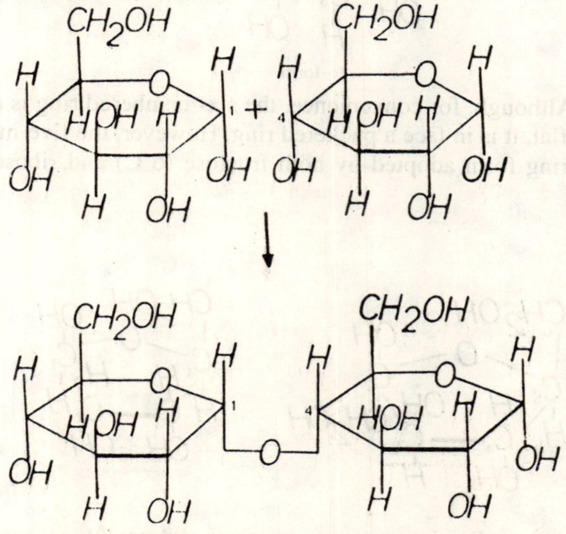

Because the left hand molecule is in the α form and the ether linkage is formed between Cl on the left molecule and C4 on the right, the linkage is called α (1 → 4). Similarly an α (1 → 2) linkage between molecules of glucose and fructose, gives sucrose–

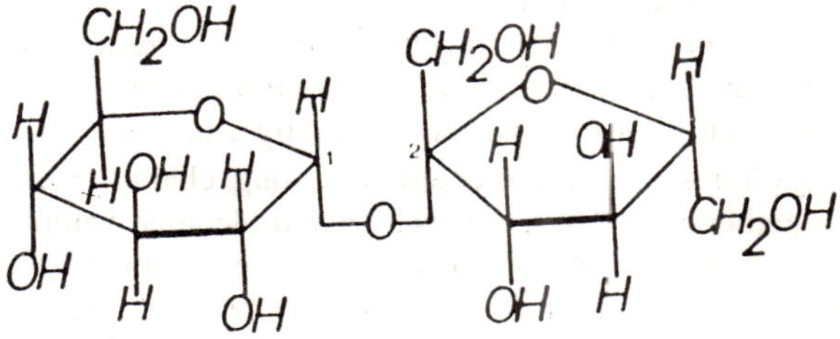

These ether or *glycosidic linkages* between rings can be broken by enzymes or by *acid hydrolysis* to give back the original monosaccharide molecules.

From this brief study of the chemistry of disaccharides, two important conclusions can be drawn :

(a) in ring closure and in the glycosidic linkage we have important exceptions to the rule that ethers are unreactive, and

(b) hydrolysis reactions are the reverse of condensation reactions in carbohydrates (and in lipids and proteins), e.g.

$$\text{glucose} + \text{glucose} \underset{\text{hydrolysis}}{\overset{\text{condensation}}{\rightleftharpoons}} \text{maltose} + H_2O$$

Sucrose, the substance we normally call sugar (cane sugar), is the most commonly occurring disaccharide and is the form in which photosynthate is transported in the phloem of higher plants. Maltose occurs in germinating seeds as a breakdown product of starch. Lactose (glucose + galactose) is a constituent of milk.

Polysaccharides

The condensation of monosaccharides can proceed further than disaccharides to give polysaccharidess complex, long-chain, macromolecules, often highly branched. The condensation of glucose molecules alone can give four different polysaccharides which are of great interest in biochemistry :

(a) Starch consists of about 25% amylose and 75% amylopectin (in cereal grain starch), where Amylose macromolecules are long unbranched chains of glucose unites bonded together by α (1–4) linkages, as in maltose, to give molecular weights of 4000 to 50 000 ; and *Amylopectin* macromolecules can be thought of as branching networks of amylose chains linked together by α (1–6) and α (1–4) glycosidic linkages. They have MW values greater than 500 000.

Starch, which gives a characteristic blue/black colour with iodine, is the form in which carbohydrate is stored in most plant species, particularly in seeds and tubers. Exceptions include certain members of the Compositae and Gramineae which store Inulin (a straight-chain polymer of 30-35 molecules of fructose terminated by a sucrose unit).

(b) Cellulose macromolecules are long chains of glucose units bonded together by β(1-4) linkages to give molecular weights of 100 000 to 500 000. Plant cell walls are made of cellulose, impregnated with other materials such as lignin. Important cellulose products include paper and cotton.

Cellulose is much more resistant to hydrolysis than is starch, and obviously much less digestible by humans and livestock ; indeed, ruminants rely on bacteria in the rumen to break down cellulose. Since, apart from differences in chain length, cellulose and starch differ principally in the configuration of the glycosidic linkage between glucose units (i.e. α (1–4) compared with β (1–4), the higher resistance to hydrolysis in cellulose must be due to the greater chemical stability of the β (1–4) linkage.

(c) Dextrins are formed when some of the glycosidic linkages in starch are broken by acid hydrolysis to give macromolecules similar to amylose and amylopectin but of lower molecular weight. Dextrins are used in glues and as food additives ('thickening' agents).

(d) Glycogen is the form in which carbohydrate is stored by animals. Its structure is very similar to amylopectin but with molecular weights of several million.

In nature, there exists a wide range of polysaccharides of glucose and other monosaccharides, commonly in association with substituted molecules, like amino sugars (e.g. gums and pectins in plants ; agar from seaweed ; chitin in the exoskeleton of insects, and lubricant, such as hyaluronic acid, in vertebrate joints). Due to the presence of many hydroxyl groups, polysaccharides are very hydrophilic, those with lower molecular weights (e.g. dextrins) forming colloidal systems with water.

In the brewing of beer, the starch in cereal grains is hydrolysed to maltose by enzymes present in germinating grain (malt). The brew is then inoculated with yeast which transforms maltose to ethanol by fermentation :

$$C_6H_{12}O_6 \rightarrow 2C_2H_5OH + 2CO_2 \uparrow.$$

As pointed out in section 13.1, care must be exercised to avoid the contamination of the fermentation mixture with micro-organisms causing the oxidation of ethanol to ethanol acid (acetic acid). Spirits can be obtained by distillation of dilute solutions of alcohol (beer, wine, etc.)

CARBOXYLIC ACIDS AND LIPIDS

Carboxylic acid molecules contain a carbonyl groups and a hydroxyl group attached to the same carbon atom at the end of a hydrocarbon chain, e.g.

$$CH_3-\overset{\overset{\displaystyle O}{\|}}{C}-OH$$

Ethanol acid
(acetic acid)

Using the systematic method of naming, we delete the final -e in the name of the corresponding alkane and substitute -oic acid.

Thus CH_4 (Methane) gives $H-\overset{\overset{\displaystyle O}{\|}}{C}-OH$ Methanoic acid (formic acid)

$$CH_3-\overset{\overset{\displaystyle CH_3}{|}}{CH}-CH_2-CH_2-CH_3 \text{ gives } CH_3-\overset{\overset{\displaystyle CH_3}{|}}{CH}-CH_2-CH_2-CO_2H$$

2–methyl-pentane 4–methyl-pentanoicacid

where the carboxylic-acid-group-carbon is always numbered Cl. Note also that non-systematic names are common especially in naming acid derivatives like esters.

The carbon atom next to the carboxylic acid group may be referred to as the α carbon and the next, the β carbon :

$$CH_3-\overset{\beta}{CH_2}-\overset{\alpha}{CH_2}-CO_2H$$

Thus it is customary to call acids of the following general structure :

$$R-\overset{\overset{\displaystyle OH}{|}}{CH}-CH_2-CO_2H$$

β-hydroxy acids. In addition, the term *acyl* group is used to indicate

$$R-\overset{\overset{\displaystyle O}{\|}}{C}-$$

where R indicates an alkyl group.

Table 16.3. Properties of some carboxylic acids.

Name	*Formula*	*Mole-cular weight*	*m.p. °C*	*b.p. °C*	*Water solubility (g 100 ml⁻¹)*	*Name of acyl group*
Methanoic (Formic) acid	HCO_2H	46	8	101	*	Formyl
Ethanolic (Acetic) acid	CH_3CO_2H	60	17	118	*	Acetyl
Propanoic (Propionic) acid	$C_2H_5CO_2H$	74	-21	141	*	Propionyl
Butanoic (Butyric) acid	$C_3H_7CO_2H$	88	-4	164	*	–
Pentanoic (Valeric) acid	$C_4H_9CO_2H$	102	-34	186	4.97	–
Hexanoic acid (Caproic)	$C_5H_{11}CO_2H$	116	-2	205	1.08	–
Decanoic acid (Capric)	$C_9H_{19}CO_2H$	172	32	270	0.015	–
Hexadecanoic (Palmitic) acid	$C_{15}H_{31}CO_2H$	256	63	390	0.0007	–
Octadecanoic (Stearic) acid	$C_{17}H_{35}CO_2H$	284	72	360	0.0003	–

* Soluble in all proportions.

Properties and Reactions of Carboxylic Acids

Since hydrogen bonds can form between two carboxylic acid groups and between carboxylic acid molecules and water, the lower molecular weight acids have relatively high melting and boiling points and high water solubilities (those up to decanoic are liquids, Table 16.3). However, as in other homologous series the hydrophobic nature of the hydrocarbon chain increasingly dominates the physical properties of the acids as the molecular weight rises. Several of the lower acids have strong odours, e.g. Butanoic smells like rancid butter, whereas C_6 to C_{10} acids are said to smell like goats.

1. *Acid/base Properties*

The highly electronegative oxygen atoms in the hydroxyl and carbonyl functional groups attract electrons strongly away from the hydroxyl hydrogen atom, i.e.

$$R-\overset{\overset{\displaystyle O}{\|}}{C}-\overset{\overset{e}{\leftarrow}}{O}-H$$

so that the hydrogen atom is not tightly held and tends to ionize in water :

$$R-\overset{\overset{\displaystyle O}{\|}}{C}-O-H+H_2O \rightarrow R-\overset{\overset{\displaystyle O}{\|}}{C}-O^-+H_3O^+$$

giving *carboxylate ions* (formate, acetate, butyrate, stearate, etc.) whose chemical stability is solution also favours the ionization. Consequently, carboxylic acids are weak acids whose degree of dissociation varies according to the following K_a values :

Methanoic acid	1.8×10^{-4}
Ethanoic acid	1.8×10^{-5}
Octanoic acid	1.3×10^{-5}

The neutralization of carboxylic acids by inorganic bases gives rise to a range of salts, many of which are useful ; for example, sodium palmitate and sodium stearate are soaps, whereas lead acetate (sugar of lead) is a rather hazardous pain-killer.

2. *Esterification and Hydrolysis of Esters*

Alcohols react with carboxylic acids to give esters which have the general formula :

$$R^1-\overset{\overset{\displaystyle O}{\|}}{C}-OR$$

For example $C_2H_5OH + CH_3CO_2H \rightarrow CH_3\overset{\overset{\displaystyle O}{\|}}{C}-OC_2H_5 + H_2O$

Ethyl acetate

$$C_6H_4(CO_2H)(OH) + CH_3CO_2H \longrightarrow C_6H_4(CO_2H)(O\overset{\overset{\displaystyle O}{\|}}{C}\text{-}CH_3) + H_2O$$

Salicylic acid

Acetyl salicylic (Aspirin) acid

Esters are important as flavouring materials, solvents, drugs, perfumes and as synthetic textiles (e.g. polyester fibres). Many have very pleasant smells which help in their identification. They can be hydrolysed back to the original acid and alcohol under alkaline conditions. In general :

$$\text{Alcohol} + \text{acid} \underset{\substack{\text{Hydrolysis} \\ \text{high pH}}}{\overset{\substack{\text{low pH} \\ \text{Esterification}}}{\rightleftharpoons}} \text{Ester} + H_2O$$

In biochemistry there are two groups of highly important complex esters–the triglyceride lipids and 'high energy' esters.

(a) *The Triglyceride Lipids* : Lipids are defined as substances, originating from living organisms, which are soluble in hydrophobic solvents (e.g. ether, chloroform, benzene) but insoluble in water. Lipids are important as energy stores in seeds and in the fat deposits of animals, but they have many other biochemical roles to play, for example, in determining the hydrophobic properties of cell membranes and in the electrical insulation of nerves. In spite of this rather loose definition, the majority of lipids belong to the same family, the triglycerides (carboxylic acid esters of glycerol), whose generalized structure is :

$$\begin{array}{l} CH_2 - O - \overset{\overset{\displaystyle O}{\|}}{C} - R_1 \\ | \\ CH_2 - O - \overset{\overset{\displaystyle O}{\|}}{C} - R_2 \\ | \\ CH_2 - O - \overset{\overset{\displaystyle O}{\|}}{C} - R_3 \end{array}$$

(R_1, R_2 and R_3 may be the same, or different groups)

where the R groups are straight alkane or alkene chains, usually of at least ten carbon atoms. Triglycerides which yield unsaturated or short-chain carboxylic ('fatty') acids on hydrolysis tend to be liquids at room temperature and are, therefore, called oils (e.g. many vegetable oils—sunflower, cottonseed, etc.), whereas those yielding saturated 'fatty acids' are solids at room temperature and are termed *fats* (e.g. various animal fats). (Note that care should be exercised to avoid confusion between the terms lipid, fat and oil ; the term fat, in particular, is commonly employed in the place of lipid.)

Table. 16.4. The most widely distributed carboxylic acids occurring in the triglycerides of plants and animals.

Acid	*Number of Carbon atoms*	*Formula*
Oleic (Unsaturated)	18	$CH_3 - (CH_2)_7 - CH = CH - (CH_2)_7 - CO_2H$
Linoleic*	18	$CH_3 - (CH_2)_4 - CH = CH - CH_2 - CH = CH - (CH_2)_7 - CO_2H$
Linolenic*	18	$CH_3 - CH_2 - CH = CH - CH_2 - CH = CH - CH_2 - CH = CH - (CH_2)_7 - CO_2H$
Lauric (Saturated)	12	$CH_3 - (CH_2)_{10} - CO_2H$
Palmitic	16	$CH_3 - (CH_2)_{14} - CO_2H$
Stearic	18	$CH_3 - (CH_2)_{16} - CO_2H$

*Essential fatty acids which must be provided in the diet of vertebrates.

Although the range of possible combinations of carboxylic acids with glycerol indicated by the general structure is very large, the majority of lipids from plants and animals contain combinations of only six carboxylic acids (Table 16.4). The 'hardening' of vegetable oils to fats is discussed in Chapter 15 as are the techniques for determining the degree of unsaturation of an oil sample. The alkaline hydrolysis (or saponification) of saturated triglycerides yields glycerol and soaps :

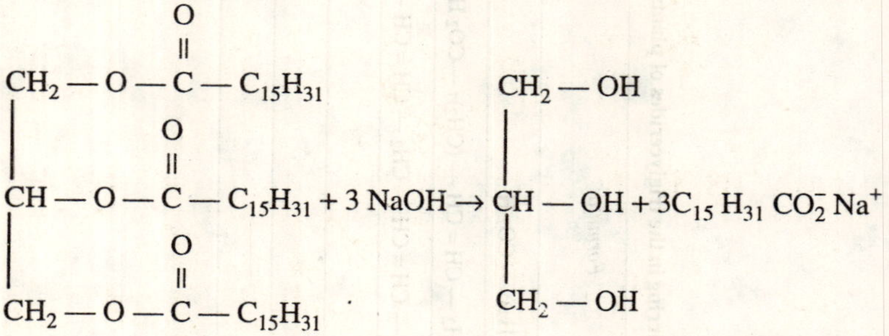

$$\begin{array}{l} CH_2-O-\overset{\overset{\large O}{\|}}{C}-C_{15}H_{31} \\ | \\ CH-O-\overset{\overset{\large O}{\|}}{C}-C_{15}H_{31} \\ | \\ CH_2-O-\overset{\overset{\large O}{\|}}{C}-C_{15}H_{31} \end{array} + 3\,NaOH \rightarrow \begin{array}{l} CH_2-OH \\ | \\ CH-OH \\ | \\ CH_2-OH \end{array} + 3C_{15}H_{31}CO_2^-Na^+$$

Sodium palmitate.

(b) *High Energy Esters* : 'High energy compounds', which can store chemical (potential) energy and release it when required, are critically important in the energy metabolism of living organisms. Examples include several esters, whose hydrolysis reactions are highly exergonic, such as :

Acyl phosphates

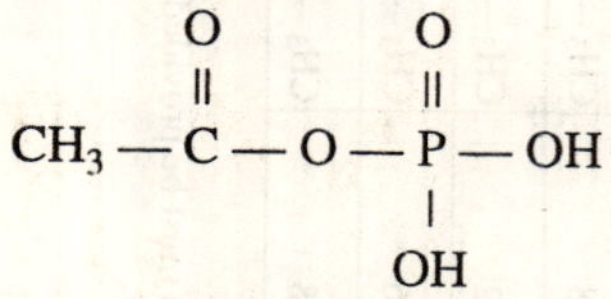

$$CH_3-\overset{\overset{\large O}{\|}}{C}-O-\underset{\underset{\large OH}{|}}{\overset{\overset{\large O}{\|}}{P}}-OH$$

(where phosphoric acid is acting as an alcohol)

Acetyl phosphate

Thioesters

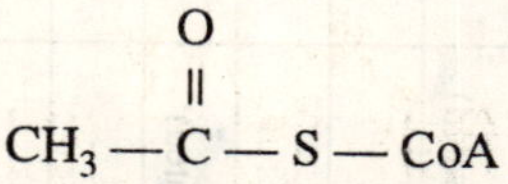

$$CH_3-\overset{\overset{\large O}{\|}}{C}-S-CoA$$

Acetyl Coenzyme A

3. *Decarboxylation Reactions*

Simple carboxylic acids do not readily lose the carboxylic acid group in chemical reactions. However, the decarboxylation of β-keto acids is one step in the metabolism of lipids, e.g.

$$CH_3-\overset{\overset{\displaystyle O}{\|}}{C}CH_2-\overset{\overset{\displaystyle O}{\|}}{C}-OH \rightarrow CH_3-\overset{\overset{\displaystyle O}{\|}}{C}CH_3+CO_2$$

4. *The Reactions of Methanoic Acid*

Methanoic acid is unique amongst carboxylic acids since it contains an aldehyde as well as a carboxylic acid group :

$$H-\overset{\overset{\displaystyle O}{\|}}{C}-OH$$

As a result, methanoic acid exhibits the reactions of both groups. For examples, it is a weak acid but also gives positive Fehling's and Tollen's tests.

Important Carboxylic Acids, Their Occurrence and Use

Methanoic (Formic) Acid is responsible for the painfulness of ant bites and nettle stings. It is used for removing hair from hides, in dyeing and in making plastics.

Ethanoic (Acetic) Acid is used in dyeing, the manufacture of textiles and as a solvent. Vinegar is 4% ethanoic acid. The pure acid is called 'glacial acetic acid' since it freezes at 17° C.

Butanoic (Butyric) Acid is partly responsible for the taste of butter. It is added to margarine to make it more palatable.

Long-chain Acids are used in soaps, candles, floor wax, shoe polish, etc.

Other carboxylic acids of importance in biochemistry include :

$$\begin{array}{l} \quad\; CH_2-CO_2H \\ \quad\;\; | \\ HO-C \\ \quad\;\; | \\ \quad\; CH_2-CO_2H \end{array}$$

Citric acid

$$\begin{array}{l} CH_2-CO_2H \\ | \\ CH_2-CO_2H \end{array}$$

Succinic acid

$$CH_3-\overset{\overset{\displaystyle O}{\|}}{C}-CO_2H$$

Pyuvic acid

and

$$\begin{array}{l} CH-CO_2H \\ \| \\ CH-CO_2H \end{array}$$

Fumaric acid

which take pat in the Krebs Cycle ;

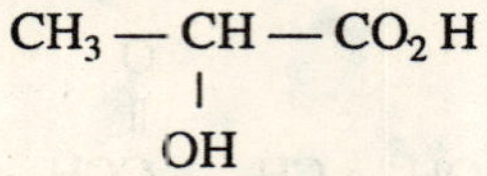

$$CH_3 — CH — CO_2 H$$
$$|$$
$$OH$$

Lactic acid

which accumulates in seeds and in muscles during anaerobic metabolism ;

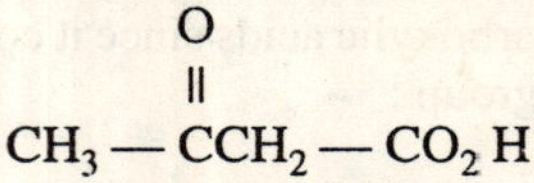

$$CH_3 — \overset{\overset{O}{||}}{C}CH_2 — CO_2 H \qquad C_3 — \overset{\overset{OH}{|}}{CH} — CH_2 — CO_2 H$$

Acetoacetic acid 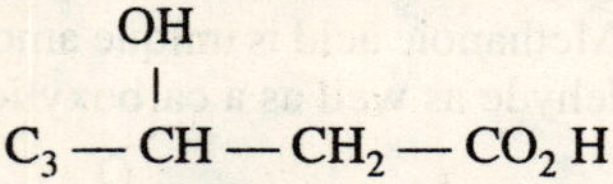   β–hydroxy–hutyricacid

which together with acetone, constitute the 'Ketone bodies' which accumulate in the blood under conditions of rapid fast degradation (e.g. in acute form in diabetes) ;

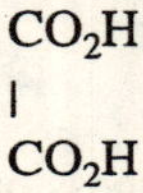

$$CO_2H$$
$$|$$
$$CO_2H$$

Oxalic acid

$$OH$$
$$|$$
$$CH—CO_2H$$
$$|$$
$$CH—CO_2H$$
$$|$$
$$OH$$

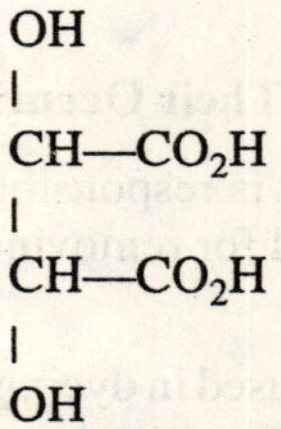

Tartaric acid

which occur at high concentrations in the leaves and fruits of certain plant species.

CHAPTER 17

Organic Compounds Containing Nitrogen, Sulphur and Phosphorous

Of the three major groups of biochemical substances, *carbohydrates* and *lipids* are composed primarily of oxygen-containing organic molecules, but in order to understand the properties of proteins, it is also necessary to study the chemistry of nitrogen- and sulphur-containing functional groups. In this chapter we shall also consider some phosphorous-containing compounds which are important in heredity and in energy metabolism.

THE AMINES AND AMIDES

The amines are aliphatic and aromatic compounds containing one or more amino (--NH_2, –NHR, etc.) groups. They can be considered as derivatives of ammonia, with substitution of one, two or three hydrogen atoms by alkyl (or aryl) groups giving primary, secondary or tertiary amines :

NH_3	RNH_2	R_2NH	R_3N
Ammonia	Primary Amine	Secondary Amine	Tertiary Amine

e.g.	CH_3NH_2	$(CH_3)_2NH$	$(CH_3)_3N$
	Methylamine	Dimethylamine	Trimethylamine.

The bonding of alkyl groups to the amine nitrogen atom can go one step further to give *quaternary* ammonium ions (analogous to the ammonia ion, NH_4^+) ; for example, choline (a component of the complex phospholipid sphingomyelin from brain and muscles.

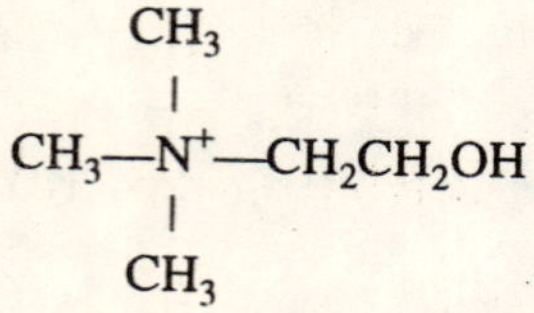

$$CH_3-\overset{\overset{CH_3}{|}}{\underset{\underset{CH_3}{|}}{N^+}}-CH_2CH_2OH$$

As we can see from the examples above, simple amines are named like ethers, the attached groups being listed, followed by the name -amine. More complicated molecules can be named systematically, e.g.

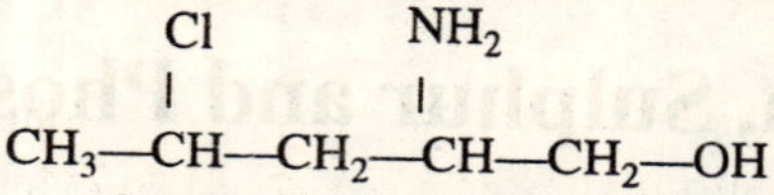

$$CH_3-\overset{\overset{Cl}{|}}{CH}-CH_2-\overset{\overset{NH_2}{|}}{CH}-CH_2-OH$$

2–amino–4–chloro–1–pentanol.

The simplest aromatic amine, amino-benzene, is normally called *Aniline*, but more complex aromatic amines are most conveniently named using the systematic method outlined in 12.4, for example—

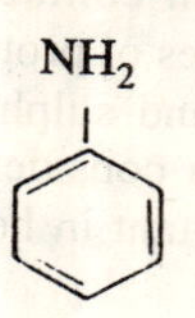

NH_2 CH_3 NH_2

NH_2

Anline 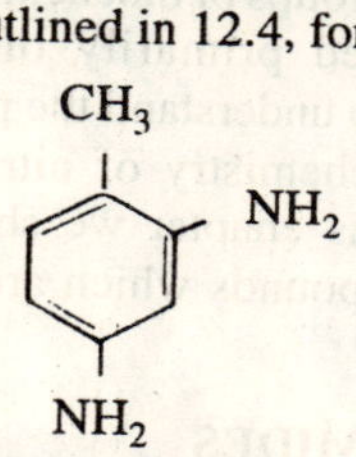 2,4 – diamino toluene

Table 17.1. Properties of some simple amines.

Name	*Formula*	*Mole-cular weight*	*m.p.°C*	*b.p.°C*	*Water solubility (g 100 ml-1)*	K_b
Ammonia	NH_3	17	– 78	– 33	89·9	$1{\cdot}8 \times 10^{-5}$
Methylamine	CH_3NH_2	31	– 94	– 6	v. soluble	$4{\cdot}4 \times 10^{-4}$
Ethylamine	$C_2H_5NH_2$	45	– 81	17	*	$5{\cdot}6 \times 10^{-4}$
Aniline	$C_6H_5NH_2$	93	–6	184	3·7	$3{\cdot}8 \times 10^{-10}$

* Soluble in all proportions.

Properties and Reactions of Amines

In general, low molecular weight amines are strong-smelling gases (methyulamine) or volatile liquids (ethylamine) whose high water solubilities are due to hydrogen bonding between the amino group and water.

1. *Acid/Base Properties*

Like ammonia, amines are weak bases, for example, in the ionization of methylamine in water (giving a quaternary ammonia ion) :

$$CH_3NH_2 + H_2O \rightleftharpoons CH_3NH_3^+ + OH^-,$$

$$K_b = \frac{[CH_3NH_3+][OH^-]}{[CH_3NH_2]} = 4.4 \times 10^{-4}$$

showing that methylamine is a weaker base than sodium hydroxide ($K_b > 1$) but stronger than ammonia ($K_b = 1.8 \times 10^{-5}$) and aniline ($K_b = 3.8 \times 10^{-10}$)

2. *The Decomposition of Amines (Kjeldahl Determination of Organic Nitrogen)*

When amines are digested with concentrated sulphuric acid in the presence of a Cu/Se catalyst, the nitrogen atoms are liberated in the form of ammonium sulphate. If the solution is then treated with a strong base (NaOH), the ionization of the ammonium ion is suppressed and gaseous ammonia will be evolved when the solution is heated. This is the principle of the Kjeldahl determination of organic nitrogen in which the ammonia evolved is collected and measured by acid/base titration. The Kjelahl method was wide applications in medicine, biochemistry, ecology and agriculture (for example in determining the nitrogen content of soil organic matter).

3. *The Condensation of Amines with Carboxylic Acids–The Amides*

Ammonia and amines react with carboxylic acids to give *amides.* In this type of reaction, the acid is normally in the form of an acid derivative (e.g. an ester) represented in generalized form as :

$$R\text{—}\overset{\overset{\displaystyle O}{\|}}{C}\text{—}X$$

For example :

$$\underset{\text{Ethanoic acid derivative}}{CH_3\text{—}\overset{\overset{\displaystyle O}{\|}}{C}\text{—}X} + NH_3 \rightarrow \underset{\text{Acetamide}}{CH_3\text{—}\overset{\overset{\displaystyle O}{\|}}{C}\text{—}NH_2} + HX$$

or more generally :

$$R-\overset{\overset{\displaystyle O}{\|}}{C}-X + R^1NH_2 \rightarrow R-\overset{\overset{\displaystyle O}{\|}}{C}-NHR^1 + HX$$

Amide

Amides contain the *amide linkage* :

$$-\overset{\overset{\displaystyle O}{\|}}{C}-\overset{\overset{\displaystyle H}{|}}{N}- \quad \text{or} \quad -\overset{\overset{\displaystyle O}{\|}}{C}-\overset{\overset{\displaystyle R}{|}}{N}-$$

which is called the *peptide linkage* in peptides and proteins. Amides can be hydrolysed to give back acid and amine (or ammonia) under both acid and basic conditions.

Acid Hydrolysis

e.g.

$$R-\overset{\overset{\displaystyle O}{\|}}{C}-NH_2 + HCl + H_2O \rightarrow R-\overset{\overset{\displaystyle O}{\|}}{C}-OH + NH_4^+Cl^-.$$

Base Hydrolysis

e.g.

$$R-\overset{\overset{\displaystyle O}{\|}}{C}-N(CH_3)_2 + NaOH \rightarrow R-\overset{\overset{\displaystyle O}{\|}}{C}-O^-Na^+ + (CH_3)_2NH.$$

The most important simple amide in biochemistry is *Urea*:

$$\begin{array}{c} NH_2 \\ | \\ C=O \\ | \\ NH_2 \end{array}$$

which is the chief end-product of mammalian protein metabolism and a major constituent of urine. It is a useful nitrogen fertilizer substance, yielding ammonium ions in soil and may also be used as a concentrated nitrogen source in livestock feeds.

When heated, urea forms a condensation product, *biuret*

$$NH_2-\overset{\overset{\displaystyle O}{\|}}{C}-NH_2 + NH_2-\overset{\overset{\displaystyle O}{\|}}{C}-NH_2 \rightarrow NH_2-\overset{\overset{\displaystyle O}{\|}}{C}-NH-\overset{\overset{\displaystyle O}{\|}}{C}-NH_2 + NH_3$$

Biuret

which gives a distinctive purple product with alkaline copper sulphate solution. (On heating, proteins also give a positive *Biure Test.*)

Biuret is not toxic to livestock but concentrations in fertilizer urea must be kept below 1% avoid phytotoxicity.

Some Important Amines

(1) The α-amino acids

(2) Several simple aliphatic amines account for the strong smells of fish and fish products, e.g. CH_3NH_2, $(CH_3)_2NH$, $(CH_3)_3N$, $C_2H_5NH_2$, etc. Tertiary amines, which can occur in bad silage, are toxic to ruminants.

(3) We have already seen that choline occurs in muscle tissues. Another quaternary ammonium compound acetylcholine (a derivative of choline), is a chemical transmitter of nerve impulses:

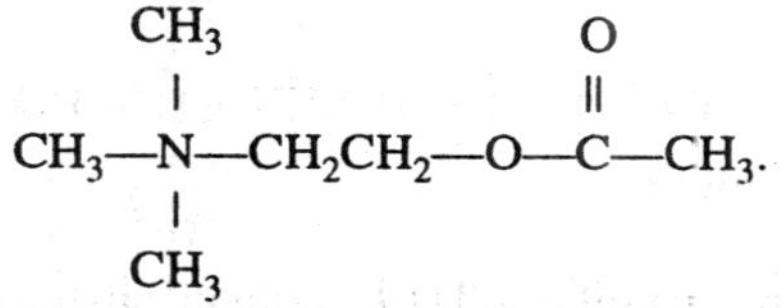

$$CH_3-\underset{\underset{CH_3}{|}}{\overset{\overset{CH_3}{|}}{N}}-CH_2CH_2-O-\overset{\overset{O}{||}}{C}-CH_3.$$

(4) The synthetic diamine EDTA (ethylene diamine tetra-acetic acid) is a chelating agent widely used in the dissolution and chemical analysis of sparingly soluble cations—

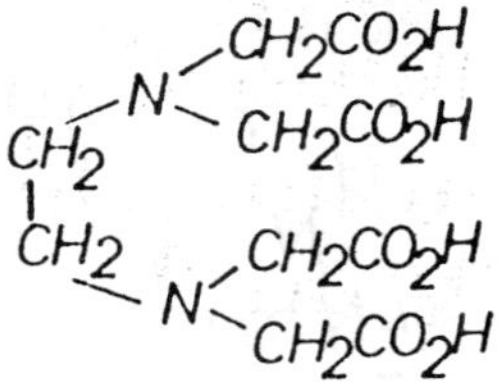

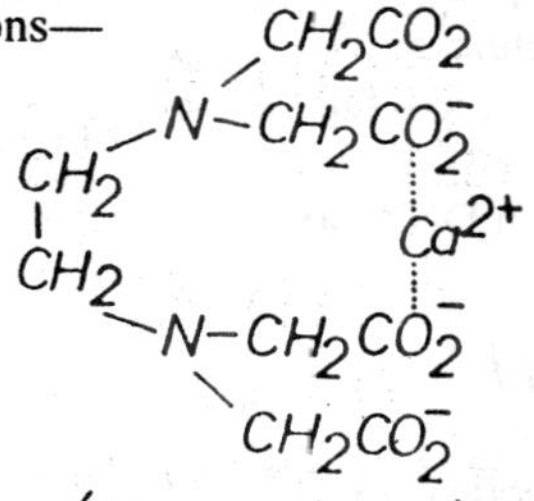

$(HO_2CCH_2)_2N-CH_2-CH_2-N(CH_2CO_2H)_2$ $\quad$ $(^-O_2CCH_2)_2N-CH_2-CH_2-N(CH_2CO_2^-)_2 \cdots Ca^{2+}$

(as an anion, chelating a Calcium cation)

(5) There exists a great variety of aromatic compounds in which one or two carbon atoms of the benzene ring have been replaced by nitrogen atoms. These *heterocyclic* aromatic amines include a large number of important compounds, for example pyrimidine and purine:

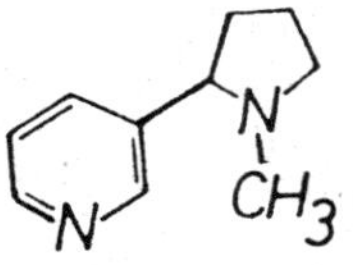

Nicotine

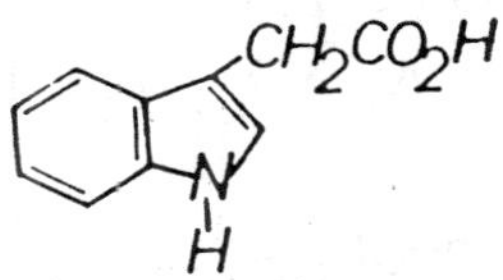

Indol-3yl-acetic acid (Auxin)
(plant growth regulator)

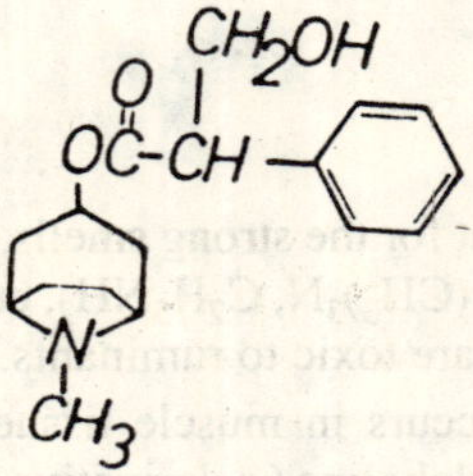

Atropine

(antidote for organophosphorus pesticide poisoning)

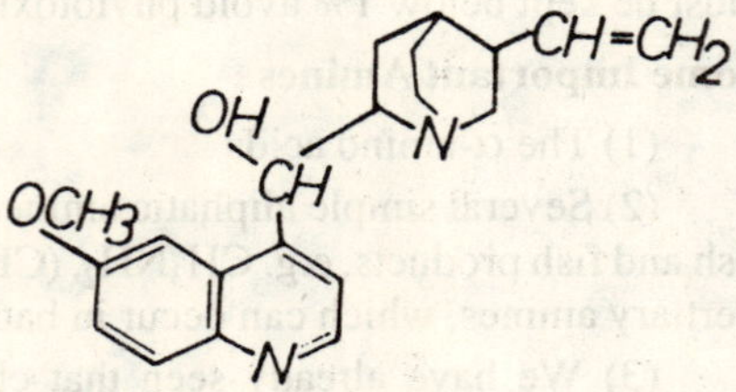

Quinine

(original drug for the treatment of malaria)

Nicotine, atropine and quinine belong to the large group of naturally occurring amine bases called alkaloids, which can be obtained from the extracts of plant tissue (notably the Umbelliferae and Solanaceae); most of them have pharmacological properties (drugs and poisons).

(6) Nucleic acids (DNA and RNA) are composed of *nucleotides* which, in turn, contain a *ribose molecule, a phosphate group and a complex amine*. A diagram of a complete nucleotide. The four amine bases which occur in DNA are—

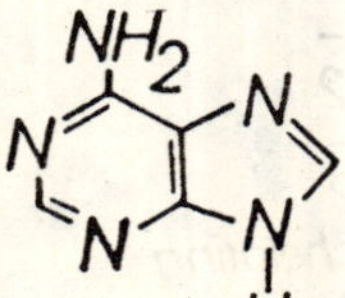

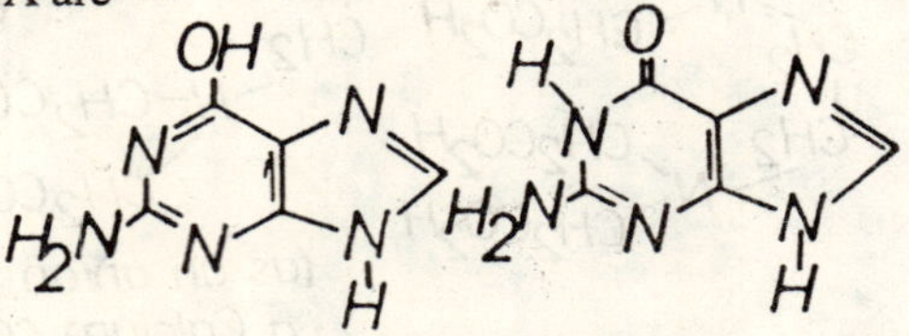

enol ⇌ keto

Adenine

Guanine

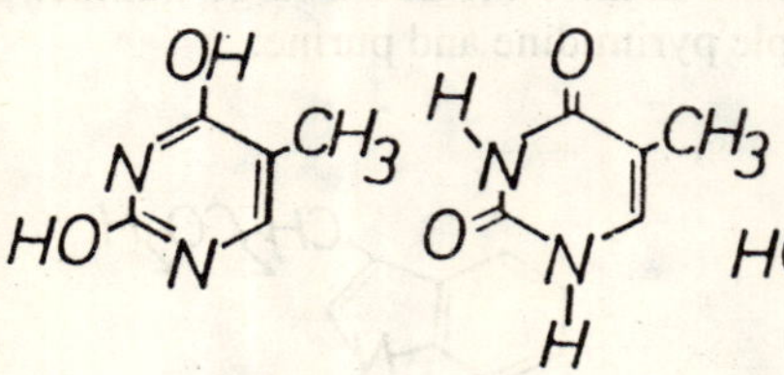

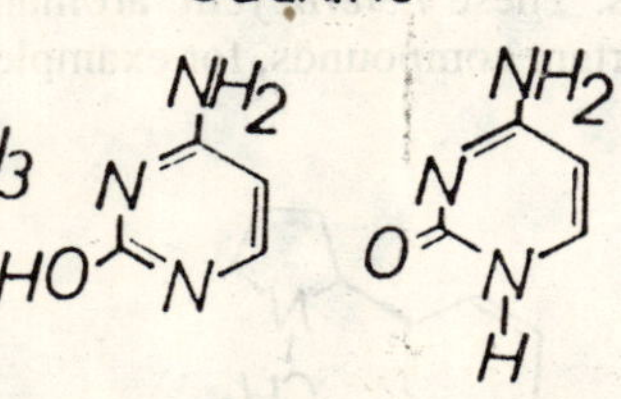

enol ⇌ keto

Thymine

enol ⇌ keto

Cytosine

Adenine and guanine are called *purine bases* because they contain the purine structure—

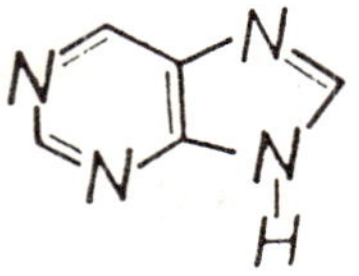

whereas thymine and cytosine are *pyrimidine bases*. When not attached to the other groups in a nucleotide, guanine, thymine and cytosine tend to take up the enol structure which is the more stable tautomer for these molecules. However, in DNA they take up the keto form since this allows adenine to form strong hydrogen bonds with thymine, and guanine with cytosine—

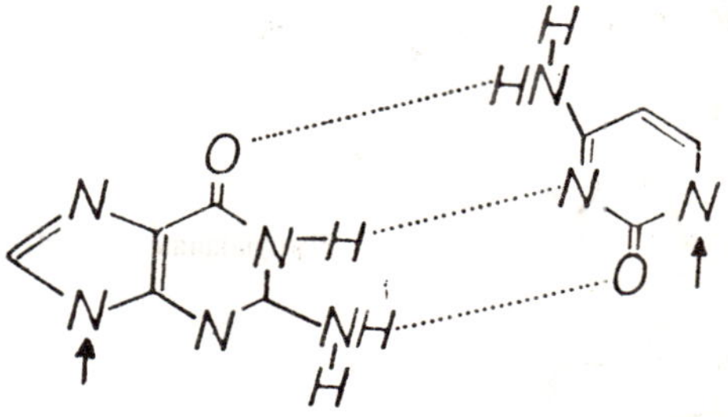

Guanine Cytosine

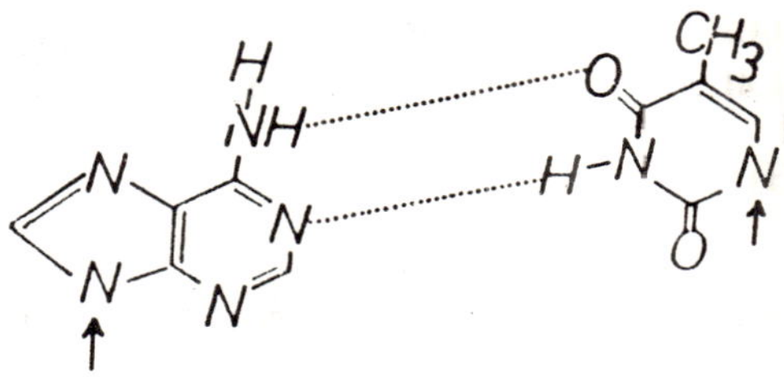

Adenine Thymine

(where the arrows indicate the points of attachment of the remainder of each nucleotide molecule). These hydrogen bonds linking the base pairs are responsible for holding together the two chains of nucleotides in the *double helix* structure (see Figure 17.3).

AMINO ACIDS, PEPTIDES AND PROTEINS

Proteins and peptides are macromolecules formed by the condensation of a large number of simpler molecules called α-amino acids, whose general formula is:

$$H_2N—\underset{|}{\overset{R}{CH}}—CO_2H$$

TABLE 17.2. Classification of α-amino acids.

Number	*R group*	*Name*	*3-letter Name*
Class 1	*R is hydrophobic*		
1	–H	Glycine	GLY
2	$-CH_3$	Alanine	ALA
3	$-CH(CH_3)_2$	Valine	VAL
4	$-CH_2CH(CH_3)_2$	Leucine	LEU
5	$-\underset{\underset{CH_3}{\vert}}{CH}CH_2CH_3$	Isolcucine	ILE
6	$-CH_2-$(phenyl ring) 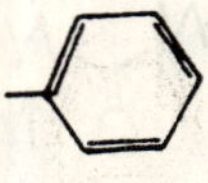	Phenylalanine	PHE
7	$-CH_2$ 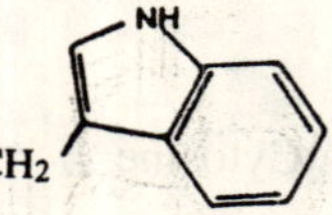 (indole ring, NH)	Tryptophan	TRP (or TRY)
Class 2	*R is hydrophilic, containing an alcohol or phenol group*		
8	$-CH_2OH$	Serine	SER
9	$-\underset{\underset{CH_3}{\vert}}{CH}OH$	Threonine	THR
10	$-CH_2-$(ring)$-OH$ 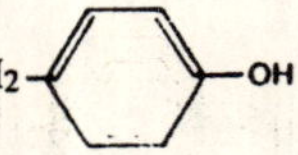	Tyrosinc	TYR
Class 3.	*R is hyrdophilic, containing a carboxylic acid or amide group*		
11	$-CH_2CO_2H$	Asparitic Acid	ASP

(*contd.*)

Number	*R group*	*Name*	*3-letter Name*
12	$-CH_2CH_2CO_2H$	Glutamic Acid	GLU
13	$-CH_2CONH_2$	Asparagine	ASN
14	$-CH_2CH_2CONH_2$	Glutamine	GLN
Class 4	*R is hydrophilic, containing an amine group**		
15	$-CH_2CH_2CH_2CH_2NH_2$	Lysine	LYS
16	$-CH_2CH_2CH_2NH-C(=NH)-NH_2$	Arginine	ARG
17	$-CH_2$– (imidazole ring: N, NH)	Histidine	HIS

*Note that although Trypotophan carries an amine group, its hydrophobic properties are determined by the attached benzene ring.

Class 5	*R contains sulphur +,*		
18	$-CH_2SH$	Cysteine	CYS
19	$-CH_2-S-S-CH_2-$	Cystine	CYS[1] or CYS-CYS
20	$-CH_2CH_2S\ CH_3$	Methionine	MET

+ Cysteine is hydrophilic and Methionine hydrophobic.

Class 6	*The imino acids++ (whole molecule shown)*		
21	(pyrrolidine ring, NH) – CO_2H	Proline	PRO
22	OH – (pyrrolidine ring, NH) – CO_2H	Hydroxyproline	HYP

++In which the amino groups forms part of a ring structure.

where the R group is attached to the α carbon of the carboxylic acid. (Compare with carbohydrate macromolecules which are composed of monosaccharide units.)

In theory, there can exist an enormous number of possible amino acids (e.g. R alkyl or aryl, branched or unbranched, saturated or unsaturated). However, in proteins from living organisms, there are only 22 common R groups, giving the 22 corresponding α-amino acids shown in Table 17.2.

Properties and Reactions of Amino Acids

1. *Acid/Base Properties*

Since amino acid molecules contain both acidic and basic groups, they behave as amphoteric substances.

2. *Transamination and Essential Amino Acids*

In their roots and leaves, photosynthetic plants reduce absorbed nitrate to ammonia, which is incorporated into the following amino acids: Aspartic Acid, Alanine, Glutamic Acid.

Other amino acids can then be synthesized from these acids by the process of *transamination.* In general terms:

$$\underset{\alpha\text{-keto acid}}{R{-}\overset{\overset{O}{\|}}{C}{-}CO_2H} + \underset{\alpha\text{-Amino acid}}{R^1{-}\overset{\overset{NH_2}{|}}{CH}{-}CO_2H} \rightarrow$$

$$\underset{\text{new }\alpha\text{-amino acid.}}{R{-}\overset{\overset{NH_2}{|}}{CH}{-}CO_2H} + R^1{-}\overset{\overset{O}{\|}}{C}{-}CO_2H$$

Plants can synthesize the entire range of amino acids but animals, in general, cannot and are dependent ultimately upon plants for certain *essential amino acids.* The range of essential amino acids varies amongst species; for example, the pig requires: arginine, histidine, isoleucine, leucine, lysine, methionine, phenylalanine, threonine, tryptophan and valine, whereas poultry require all of these and glycine in addition.

3. *Formation of Amide or Peptide Links*

In the last section, we noted that amines react with carboxylic acids to give amides, i.e.

$$RCO_2H + R^1NH_2 \rightarrow RCONHR^1.$$

Since they contain both amine and carboxylic acid groups, two amino acid molecules can condense together by this reaction to give what is called a dipeptide, e.g.

$$\underset{\text{GLY}}{H_2N{-}CH_2{-}CO_2H} + \underset{\text{ALA}}{H_2N{-}\overset{\overset{CH_3}{|}}{CH}{-}CO_2H} \rightarrow$$

$$\underset{\text{GLY-ALA (dipeptide)}}{H_2N{-}CH_2{-}\overset{\overset{O}{\|}}{C}{-}\overset{\overset{H}{|}}{N}{-}\overset{\overset{CH_3}{|}}{CH}{-}CO_2H}$$

where the amide link between the two amino acids is called a peptide link. A dipeptide is still an amino acid (although not an α-amino acid) and it can condense with a further α-amino acid to give a tripeptide, e.g. GLY-ALA-LEU. Obviously this can go on to give very long chains of amino acids; peptides containing less that 100 α-amino acid) acids are normally called polypeptides, whereas those with 1ᴖᴖ to 10 000 or more are proteins. The peptide link in peptides and proteins may be broken to give back the original α-amino acids by acid hydrolysis or by the action of enzymes.

Peptides and proteins are of enormous importance in biochemistry: as harmones controlling metabolism and physiology; as enzymes regulating metabolic reactions; as respiratory pigments transporting oxygen to sites of respiration; in the immune response and as structural elements in muscles, joints and connective tissues.

The Structure of Peptides and Proteins

The *primary structure* of peptides and proteins is simply the *order* in which the α-amino acids are assembled; for example, the peptide hormone, *Oxytocin* (released from the pituitary—causes birth contractions in the uterus) has the following primary structure:

H_2N—GLY—LEU—PRO—CYS—ASN—GLN—ILF—TYR—
CYS—CO_2II

where the ends carrying *free* amine and carboxylic acid groups are indicated by —NH_2 and —CO_2H respectively.

Over the last thirty years, the primary structures of a large number of polypeptides and smaller proteins have been elucidated; for example the hormine insulin (51 amino acids); the enzyme ribonuclease (124) and the respiratory pigment myoglobin (153) (Figure 17.2). Some idea of the variation in amino acid composition amongst proteins is given by Table 17.3.

TABLE 17.3. The distribution of nine amino acids in selected proteins (g 100 g^{-1} of protein).

Protein	*GLY*	*ALA*	*VAL*	*LEU*	*ILE*	*MET*	*PHE*	*TRP*	*LYS*
Fibroin (silk)	44	30	4	1	1	0	3	0	1
Keratin (wool)	7	4	5	11	0	1	4	2	3
Albumin (hen)	3	7	7	9	7	5	8	1	6
Haemoglobin (horse)	6	7	9	15	0	1	8	2	9
Insulin ox	4	5	8	13	3	0	8	0	3

A chain of linked amino acids does not normally remain in extended form but tends to coil up into an α-helix (Figure 17.1) which, by maximizing the number of intramolecular hydrogen bonds formed, is its most stable configuration. However, this coiling is not complete in most proteins and the *secondary structure* of the macromolecule is a measure of the extent and position of helical and extended chain sections. For example, only 70% of the chain length of myoglobin is helical (Figure 17.2). The other important type of secondary structure is the pleated sheet which occurs in fibrous proteins like silk and keratin (hair and nails).

In most proteins, the helical chain folds up into an apparently irregular, but exactly defined, three-dimensional shape which is essential for the function of the protein (as an enzyme, repiratory pigment or in structural tissues like muscles and tendons). This irregular folding of the helical chain is the *tertiary structure* of the protein and is caused by

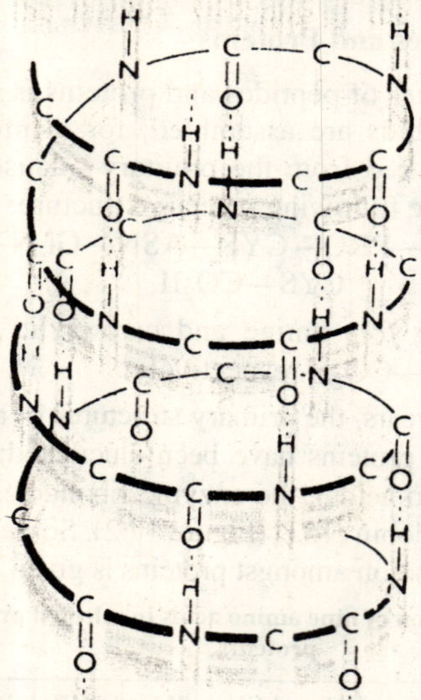

Fig.17.1. The α-helical structure of peptide and protein chains, held together by intramolecular hydrogen bonds.

interactions between functional groups attached to different parts of the chain (Table 17.2) i.e.

(a) attraction between hydrophobic groups;

(b) hydrogen bonding between hydrophilic groups, and

(c) disulphide bonds between cysteine molecules .

For example, Figure 17.2 illustrate the three-dimensional tertiary structure of myoglobin, a respiratory pigment involved in the storage of oxygen in tissues (especially in diving organisms like whales). In common with a number of respiratory pigments and enzymes, myoglobin contains a non-protein component, in this case an oxygen-binding haem group, embedded in the tertiary structure. Such inclusions are termed prosthetic groups.

In certain protein molecules, which are made up of a number of separate helical chains bound together by functional group interactions rather than peptide linkages, the three dimensional arrangements of the chains make up the *quaternary* structure of the molecule. For example, each haemoglobin molecule is composed of two 'α chains' and two 'β chains', each chain containing a haem prosthetic group embedded in its tertiary structure.

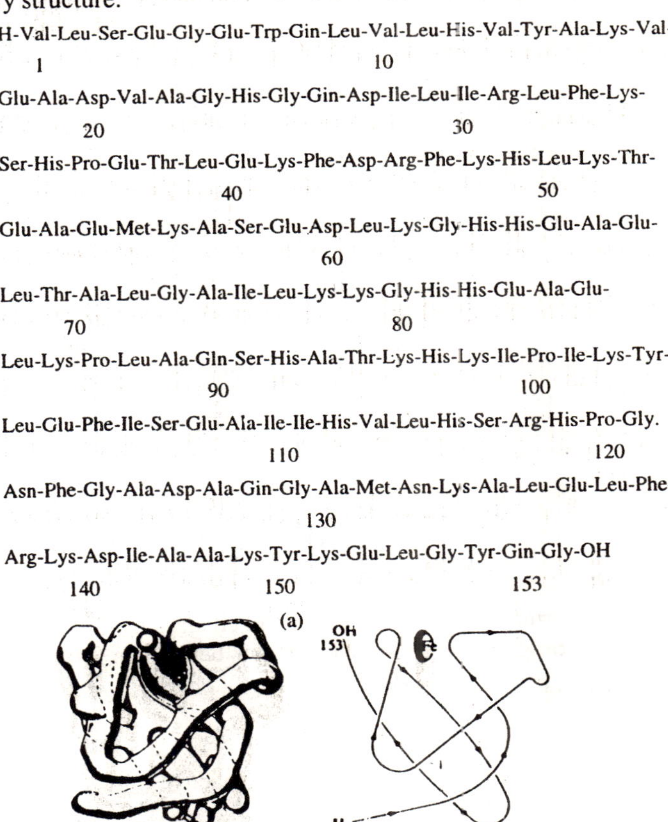

Fig.. 17.2. The primary, secondary and tertiary structure of whale myoglobin.

Factors such as heat, radiation or light, which cause changes in the secondary, tertiary or quaternary structures of a protein, without breaking peptide linkages between amino acids, are said to cause *denaturation* of the protein. A familiar example is the denaturation of egg white on boiling.

THE THIOLS

Thiols are organic compounds containing one or more thiol (SH) groups and are sometimes called thioalcohols or mercaptans. In general, thiols are of little importance in elementary chemistry; however, because of the existence of an important thiol amino acid, cystcine:

$$\begin{array}{c} CH_2SH \\ | \\ H_2N—CH—CO_2H \end{array}$$

it is essential to have some knowledge of the chemistry of the thiol functional group.

Properties and Reactions of Thiols

1. *Acid/Base Properties*

Thiols are very weak acids, but slightly stronger than alcohols e.g.

$$C_2H_5SH + H_2O \rightleftharpoons C_2H_5S^- + H_3O^+ \quad K_a = 10^{-12}$$

Under suitable conditions, salts of thiols do occur. This property is used in cell biology, where proteins containing cysteine may be precipitated, and, therefore, inactivated, as insoluble lead salts.

2. *Oxidation to form Disulphides*

Under mild oxidizing conditions, two thiol molecules may react together to give a disulphide:

$$R—SH + HS—R \rightarrow R—S—S—R.$$

The disulphide bonds are involved in determining the tertiary structures of many proteins (e.g. insulin, and keratin in hair) by bonding together different regions of the chain of amino acids. In a simpler peptide, oxytocin, the formation of a disulphide bond changes the molecule from the inactive, straight-chain form, to the active, bent-chain form :

$$\begin{array}{l} H_2N—GLY—LEU—PRO—\underset{\displaystyle SH}{\underset{|}{CYSS}}—ASN—GLN—ILE—TYR—\underset{\displaystyle SH}{\underset{|}{CYS}}—CO_2H \end{array}$$

Inactive

```
H2N—GLY—LEU—PRO—CYS—ASN
                 |    |
                 S   GLN
                 |    |          Active
                 S   ILE
                 |    |
           HO2C—CYS—TYR
```

Here, two cysteine groups have reacted to give one cystine (Table 17.2).

The disulphide bond is also employed in the biochemical oxidizing agent, Lipoic acid:

```
CH2—CH2—CHCH2CH2CH2CH2CO2H  ⇌
|       |
S       S
  Oxidized
CH2—CH2—CH2—CHCH2CH2CH2CH2CH2CO2H
|            |
SH           SH
      Reduced
```

3. *Esterification*

Thiols can substitute for alcohols in esterification reactions, i.e.

$$R—\overset{\overset{O}{\|}}{C}—OH + R^1—SH \rightarrow R—\overset{\overset{O}{\|}}{C}—SR^1 + H_2O.$$

In particular, the complex thiol, Co-enzyme A—

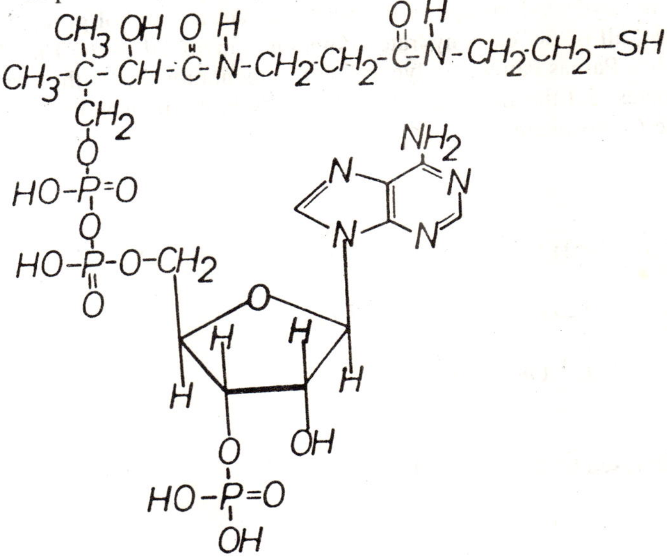

```
   CH3 OH O H            O H
    |  |  ‖ |            ‖ |
CH3-C- CH-C-N-CH2-CH2-C-N-CH2-CH2-SH
    |
    CH2
    |
    O
    |
 HO-P=O
    |
    O
    |                          NH2
 HO-P-O-CH2              (adenine)
    ‖     (ribose ring with H, H, H, H; 3'-O-PO(OH)2, 2'-OH)
    O
              O   OH
              |
          HO-P=O
              |
              OH
```

gives the thioester acetyl Co-enzyme A which participates widely in energy metabolism and biochemical syntheses.

Other thiols of biological importance include the foul-smelling butanethiol, C_4H_9SH, exuded by the skunk to repel predators. Note that the α-amino acid methionine is a thioether (Table 17.2).

ORGANIC COMPOUNDS CONTAINING PHOSPHORUS

Many compounds involved in biochemical reactions (energy and lipid metabolism) and in heredity are derivatives of phosphoric acid or one of its condensed forms (pyrophosphoric acid and triphosphoric acid):

$$\mathrm{HO{-}\overset{\overset{\large O}{\|}}{\underset{\underset{\large OH}{|}}{P}}{-}OH}$$

phosphoric acid

$$\mathrm{HO{-}\overset{\overset{\large O}{\|}}{\underset{\underset{\large OH}{|}}{P}}{-}O{-}\overset{\overset{\large O}{\|}}{\underset{\underset{\large OH}{|}}{P}}{-}OH}$$

pyro- or diphosphoric acid

$$\mathrm{HO{-}\overset{\overset{\large O}{\|}}{\underset{\underset{\large OH}{|}}{P}}{-}O{-}\overset{\overset{\large O}{\|}}{\underset{\underset{\large OH}{|}}{P}}{-}O{-}\overset{\overset{\large O}{\|}}{\underset{\underset{\large OH}{|}}{P}}{-}OH}$$

triphosphoric acid

where the hydroxyl groups of the acids can act as alcohol groups. For example, all of the carbohydrate compounds involved in the Embden-Meyerhof-Parnas Pathway (glycolysis or fermentation) are in the form of phosphates and the first step of the pathway is the phosphorylation of glucose-6-phosphate:

$$\mathrm{H{-}\overset{\overset{\large O}{\|}}{C}{-}\overset{\overset{\large OH}{|}}{\underset{\underset{\large H}{|}}{C}}{-}\overset{\overset{\large H}{|}}{\underset{\underset{\large OH}{|}}{C}}{-}\overset{\overset{\large OH}{|}}{\underset{\underset{\large H}{|}}{C}}{-}\overset{\overset{\large OH}{|}}{\underset{\underset{\large H}{|}}{C}}{-}\overset{\overset{\large O{-}\!\!\!-\!\!\!-\overset{\overset{\large O}{\|}}{\underset{\underset{\large OH}{|}}{P}}{-}OH}{|}}{\underset{\underset{\large H}{|}}{C}}{-}H}$$

catalysed by the enzyme hexokinase.

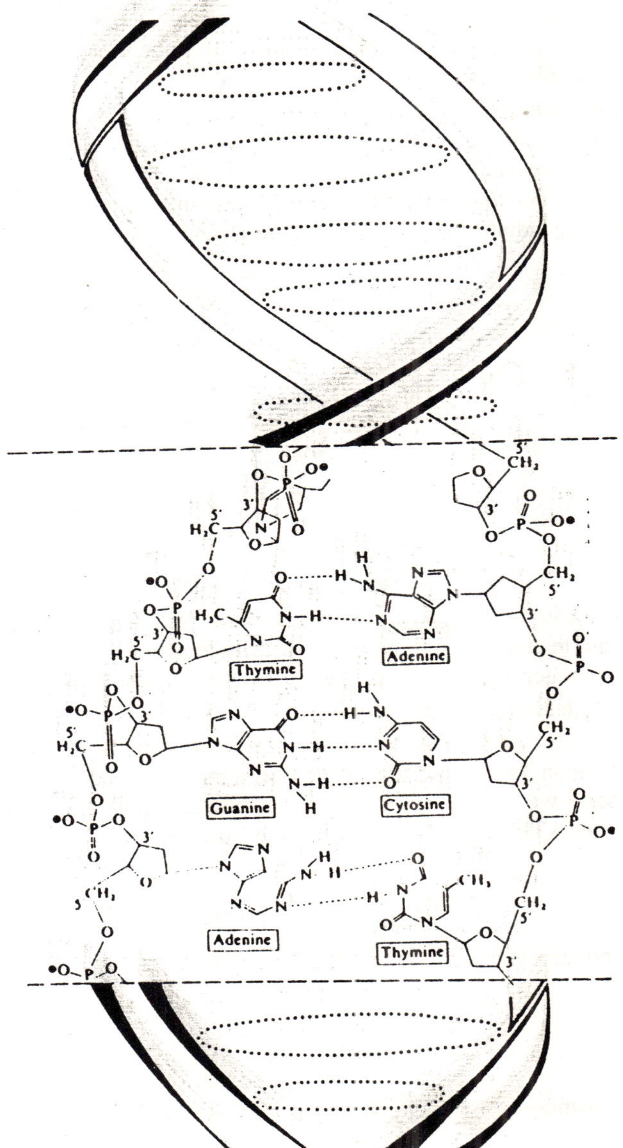

Fig.13.3. The double helix structure of DNA, held together by intermolecular hydrogen bonds.

In general, the chemistry of phosphorylated organic compounds is rather complex. However, even at an elementary level, it is essential to be able to recognize the following groups of phosphorus-containing biochemical substances.

(a) Nucleic Acids

Nucleic acids are polymers of four different nucleotides. In DNA, these are *adenosine phosphate* (contains adenine); *thymidine phosphate* (thymine); *guanosine phosphate* (guanine) and *cytidine phosphate* (cytosine), e.g. adenosine phosphate (from DNA)—

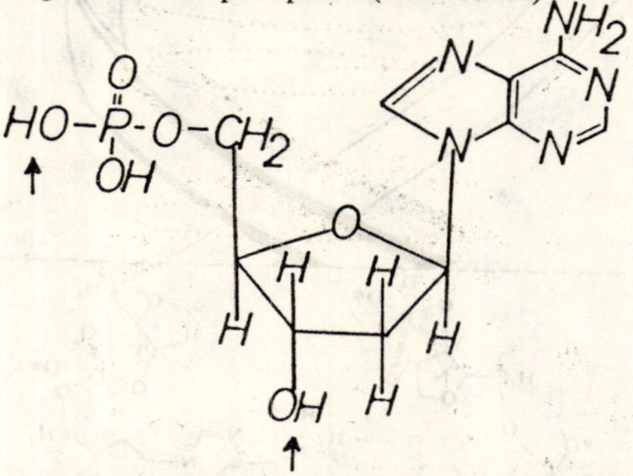

where the arrows indicate the points of attachment to adjacent nucleotides giving a single chain which pairs with a complementary single chain to give the hydrogen-bonded double helix of DNA (Figure 17.3).

(b) High Energy Compounds

Replacing the single phosphate group in adenosine phosphate (adenosine monophosphate or AMP) with a pyrophosphate or triphosphate group, we obtain adenosine diphosphate (ADP) or adenosine triphosphate (ATP). Because the hydrolysis reactions:

(1)

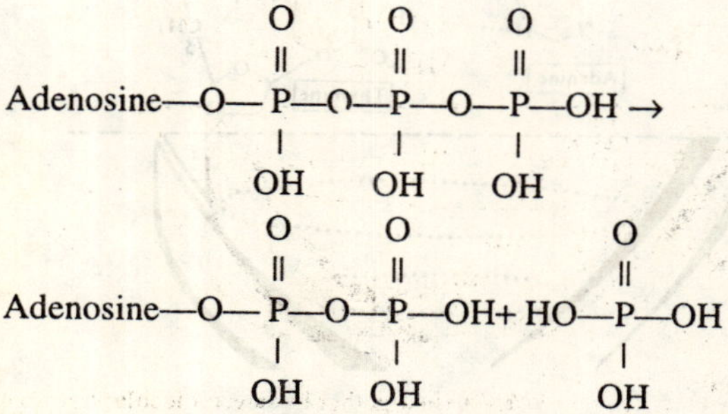

$$\text{Adenosine—O—}\overset{\overset{\displaystyle O}{\|}}{\underset{\underset{\displaystyle OH}{|}}{P}}\text{—O—}\overset{\overset{\displaystyle O}{\|}}{\underset{\underset{\displaystyle OH}{|}}{P}}\text{—O—}\overset{\overset{\displaystyle O}{\|}}{\underset{\underset{\displaystyle OH}{|}}{P}}\text{—OH} \rightarrow$$

$$\text{Adenosine—O—}\overset{\overset{\displaystyle O}{\|}}{\underset{\underset{\displaystyle OH}{|}}{P}}\text{—O—}\overset{\overset{\displaystyle O}{\|}}{\underset{\underset{\displaystyle OH}{|}}{P}}\text{—OH} + \text{HO—}\overset{\overset{\displaystyle O}{\|}}{\underset{\underset{\displaystyle OH}{|}}{P}}\text{—OH}$$

or $\quad \text{ATP} \rightarrow \text{ADP} + \text{Pi}$

(2)

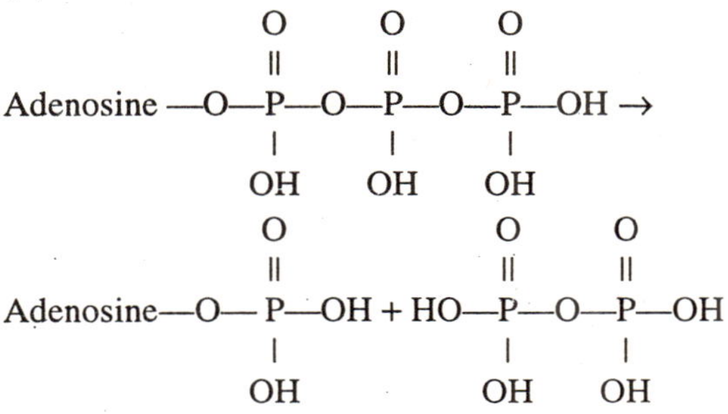

$$\begin{array}{l} \qquad\qquad\quad\;\; O \qquad\; O \qquad\; O \\ \qquad\qquad\quad\;\; \| \qquad\;\; \| \qquad\;\; \| \\ \text{Adenosine}—O—P—O—P—O—P—OH \rightarrow \\ \qquad\qquad\quad\;\; | \qquad\;\; | \qquad\;\; | \\ \qquad\qquad\quad OH \qquad OH \qquad OH \end{array}$$

$$\begin{array}{l} \qquad\qquad\quad\;\; O \qquad\qquad\qquad\quad O \qquad\; O \\ \qquad\qquad\quad\;\; \| \qquad\qquad\qquad\quad\; \| \qquad\;\; \| \\ \text{Adenosine}—O—P—OH + HO—P—O—P—OH \\ \qquad\qquad\quad\;\; | \qquad\qquad\qquad\quad\; | \qquad\;\; | \\ \qquad\qquad\quad OH \qquad\qquad\qquad OH \qquad OH \end{array}$$

or $\qquad ATP \rightarrow AMP + PPi$

are highly exergonic (2.9×10^4 J mole^{-1} for equation 1), ATP is a 'high energy compound' (section 13.6), used in metabolism to store and transfer energy. It is conventional to write ATP as:

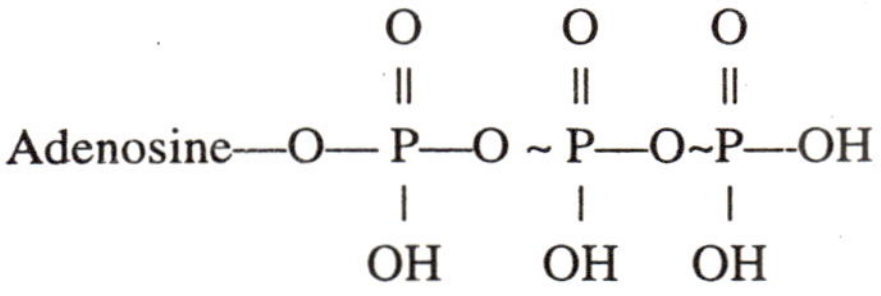

$$\begin{array}{l} \qquad\qquad\quad\;\; O \qquad\; O \qquad\; O \\ \qquad\qquad\quad\;\; \| \qquad\;\; \| \qquad\;\; \| \\ \text{Adenosine}—O—P—O \sim P—O \sim P—OH \\ \qquad\qquad\quad\;\; | \qquad\;\; | \qquad\;\; | \\ \qquad\qquad\quad OH \qquad OH \qquad OH \end{array}$$

where ~ indicates a 'high energy bond' whose hydrolysis releases a large amount of energy.

(c) Phospholipids

In addition to the simple triglycerides discussed in section 13.6, there also exist in membranes, seeds, and, particularly in brain and nerves, a variety of more complex triglycerides (phospholipids) varying in structure from the relatively phosphatides:

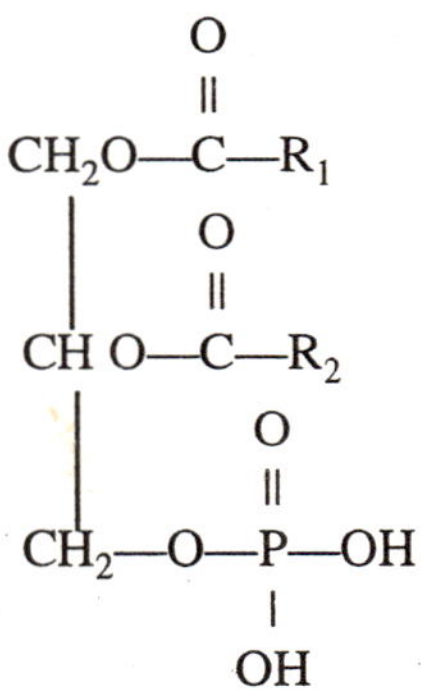

$$\begin{array}{l} \qquad\qquad\; O \\ \qquad\qquad\; \| \\ CH_2O—C—R_1 \\ |\qquad\quad\;\; O \\ |\qquad\quad\;\; \| \\ CH\;O—C—R_2 \\ |\qquad\qquad\; O \\ |\qquad\qquad\; \| \\ CH_2—O—P—OH \\ \qquad\qquad\quad | \\ \qquad\qquad\; OH \end{array}$$

to sphingomyclins, e.g.:

$$
\begin{array}{l}
\text{OH} \\
| \\
\text{CH—CH = CH—}(\text{CH}_2)_{12}\text{—CH}_3 \\
|\qquad\qquad\quad \text{O} \\
|\qquad\qquad\quad \| \\
\text{CH—NH—C—}(\text{CH}_2)_{16}\text{—CH}_3 \\
|\qquad\qquad \text{O} \\
|\qquad\qquad \| \\
\text{CH}_2\text{—O—P—O—CH}_2\text{—CH}_2\text{—N}^+(\text{CH}_3)_3 \\
\qquad\qquad\quad | \\
\qquad\qquad\quad \text{O}
\end{array}
$$

PART—III
AGRICULTURAL CHEMICALS

CHAPTER 18

Fertilizers

Crop plants (like natural plant species) require adequate supplies of between 16 and 23 chemical elements for healthy growth and high yield. These elements can be classified into four groups :

(1) H, C, O—supplied by the atmosphere and water ;

(2) N, Mg, P, S, K, Ca—the *Macronutrients* (some authorities include Fe in this group). Required in large quantities (i.e. at kg ha^{-1} levels in soils) ;

(3) B, Cl, Mn, Fe, Cu, Zn, Mo—the *Micronutrients* or *Trace Elements* required in small quantities (at ppm levels in soils) ;

(4) Na, Si, V, Co, Ni, Se, W elements involved in the growth or metabolism of certain species, e.g. Legumes require Co ; Na appears to be essential for plants using the C_4 pathway in photosynthesis.

Although several of these essential elements are abundant overall in the earth's crust Mg 2·1 %, K 2·6 %, Ca 3·6 %, Fe 5·1 %), any of the members of groups (2), (3) and (4) (above) can be deficient in soils, due to low concentrations of the element in a particular parent rock or loss during weathering and leaching of rock and soil. Since a shortage of any one essential element can cause depressions in crop yield, even if all other factors are favourable, the supply of deficient nutrients by manures and fertilizers has become an important aspect of crop production.

Until recently, farmers supplied essential nutrients to crops either by applying manure and crop residues to the soil or by various land husbandry practices (rotations, fallowing, shifting cultivation, etc.). The moisture content and macronutrient (N, P and K) levels in farm manures can vary over an enormous range according to the species age, condition and diet of the livestock involved and also to the management systems employed. Thus there is a marked contrast between liquid pig slurry (typically

< 5 % dry matter, 0·4 % N, 0·1 % P and 0·2 % K, expressed on a fresh weight basis) and poultry manure (typically 30% dry matter, 1·7 % N, 0·6 %P and 0·6 % K). Clearly, their high moisture content and variability in composition, give rise to serious practical problems in the use of manures for accurate fertilization of crops. On the other hand, manures are particularly useful in supplying adequate and balanced quantities of micronutrients, although slurry from intensive livestock enterprises can contain very high, and potentially toxic, levels of Cu. Other organic materials which have been used widely in agriculture and horticulture include sewage sludge, blood, bone meal, fish scrap, green manures, guano (sea bird manure) and compost.

Due to a large number of factors (e.g. separation of arable from livestock farming, inadequate supplies of manure, variability in the quality of manures, high transport costs and the high nutrient requirements of improved crop varieties—in particular the newer dwarf cereals), synthetic, inorganic fertilizers are used widely in intensive agriculture, sometimes in conjunction with organic manures. This chapter examines some relevant chemical characteristics of these synthetic fertilizers, concentrating on N, P and K compounds.

NITROGEN FERTILIZERS

Although nitrogen is very abundant in the atmosphere (78%), it exists in the form of very stable and unreactive diatomic molecules. Consequently, the conversion of atmospheric nitrogen into those forms which are useful to non-leguminous plants (i.e. NH_4^+ and NO_3^-) involves industrial processes requiring considerable quantities of energy (high pressure and high temperature reactions). Before the development of the Haber Process for the synthesis of ammonia about 65 years ago, farmers depended mainly upon manures, nitrogen-fixing legumes (in various rotation systems), guano (seabird manure from S. America—up to 13% N and 9% P) and Chile Nitrate, for the supply of nitrogen to crops. Since then, and particularly since 1940, synthetic nitrogen fertilizers have become increasingly important, with annual production of fertilizer from the Haber Process alone exceeding 100 million tonnes.

Chemical Properties of Nitrogen Fertilizers (Table 18.1)

Ammonium sulphate, originally a by-product of coal-gas generation, was for a long time the most popular nitrogen fertilizer, but it has now been widely superceded by ammonium nitrate (the most important compound in the UK), urea and ammonia, because of their higher nitrogen contents. Anhydrous ammonia is used particularly on highly mechanized farms in North America where it is stored as a liquid under pressure and

Table 18.1. Properties of nitrogen fertilizers.

Fertilizer	*Chemical substance(s)*	*Solubility (g 100 ml^{-1} water)*	*% N*	*Preparation or source*
Ammonium sulphate	$(NH_4)_2SO_4$	71	21	From NH_2 and H_2SO
Sodium nitrate	$NaNO_3$	92	16	Chile nitrate deposits
Ammonium nitrate	NH_4NO_3	118	35	From NH_3 and HNO_3
Nitrochalk*	NH_4NO_3 and $CaCO_3$	118 0.0014	20-26	From NH_3 and HNO_3 with added limestone
Urea	$CO(NH_2)_2$	100	47	From NH_3 and CO_2
Anhydrous ammonia	NH_3	90	82	Haber Process

*Nitrochalk is the trade name for this product in the UK ; also called Ammonium Nitrate with Lime (ANL) in the USA and Calcium Ammonium Nitrate (CAN) elsewhere.

injected into the soil or into irrigation water.

Since all nitrogen fertilizer substances are highly soluble in water (Table 18.1), they are readily available in the soil solution for uptake by plant roots. However, their high solubilities can also result in substantial losses of nitrogen by leaching if heavy rainfall follows fertilizer application. Because of this and other losses (e.g. denitrification) only 40 to 60% of applied fertilizer nitrogen can be recovered in the grain and straw of cereal crops in the UK. Higher recovery rates, 60-75% can be obtained from grassland.

A common feature of ammonium-containing fertilizers is their tendency to lower the pH of soils. There are two reasons for this ; firstly, ammonium ions bind to the soil colloids (by ion exchange reactions) and are gradually converted to nitrate by soil microbial activity :

$$NH_4^+ + 2O_2 + H_2O \rightarrow NO_3^- + 2H_3O^+ \text{ (nitrification).}$$

thus increasing the hydronium ion concentration of the soil solution. This reaction also leads to the second effect : nitrate ions (from the fertilizer and from nitrification) tend to be leached out of the soil in drainage water. However, the fertilizer cations (NH_4^+) have disappeared during nitrification and therefore, in order to maintain electroneutrality, metal cations (especially K^+ and Ca^{2+}) must accompany the nitrate ions and are lost from the soil to the ground-water. Thus the natural tendency for well-drained soils to become more acidic with time, is intensified by the application of ammonium fertilizers. This effect can largely be avoided by the addition of ground limestone to the fertilizer, as in Nitrochalk (Table 18.1), or by using nitrate fertilizers only.

Although urea releases ammonium ions in contact with soil, by the reaction :

$$CO(NH_2)_2 + 2H_2O \rightarrow (NH_4)_2CO_3$$

this does not lead to acidification because of the low solubility of carbonate ions. However, care must be exercised when applying urea and ammonia to crops since high concentrations in soils can be phytotoxic.

Care must also be exercised in the storage and handling of nitrogen fertilizers ; ammonium nitrate is particularly troublesome since it is hygroscopic (absorbs moisture from the atmosphere) and also a strong, and potentially explosive, oxidizing agent. Some fertilizers provide significant quantities of other essential nutrients. For example, ammonium sulphate contains 24% S, whereas Chile Nitrate (from natural mineral deposits in South America) provides varying quantities of trace elements.

PHOSPHORUS FERTILIZERS

Plants absorb phosphorus from the soil in the form of phosphate ions ($H_2PO_4^-$ or HPO_4^{2-}). The earliest phosphorus fertilizers were ground-up bones or rock phosphate, both of which contain the rather insoluble phosphate mineral, apatite (Table 18.2). However, around 1840, it was discovered that treating bones or rock phosphate with sulphuric acid gave much more soluble phosphate which were, therefore, more readily available for plant uptake. The product from rock phosphate was called *superphosphate* (later becoming *single* or *ordinary* superphosphate) but this has largely been superceded by concentrated, *double* or *triple* superphosphate (different names for the same product) whose higher phosphorus content (Table 18.2) is due to the substitution of phosphoric acid for sulphuric acid. The use of bones as fertilizers is now restricted to horticultural applications of bone meal.

Chemical Properties of Phosphorus Fertilizers (Table 18.2)

In contrast to nitrogen fertilizer compounds, even the more soluble superphosphate fertilizers are only sparingly soluble in water. Consequently, the 'available' phosphorus in soils includes phosphate ions in the soil solution *and* those 'labile' solid phosphate which will dissolve readily to replace phosphate ions which have been absorbed by the crop. The aim of phosphorus fertilization is, therefore, to add to

Table 18.2. Properties of phosphorus fertilizers.

Fertilizer	*Chemical substance*	*Solubility (g 100 m^{-1} water)*	*% P*	*% of P in fertilizer available to plants*
*Rock phosphate	Apatite $Ca_{10}(PO_4, CO_3)_6(F, Cl, OHs)_2$	sparingly soluble	6—18	14—65
Ordinary/single superphosphate	Monocalcium phosphate $Ca(H_2PO_4)_2$	1·8		
	+		8—10	97—100
	Gypsum $CaSO_4$	0·209		
Concentrated/triple superphosphate	$Ca(H_2PO_4)_2$	1·8	19—23	96—99
† Basic slag	Complex mixture of Ca silicates and phosphates	variable (sparingly soluble)	3—10	62—94

*Obtained from natural deposits.

†By-product of the blast furnace—

the soil, phosphorus compounds which will maintain soil solution concentrations of phosphate at levels ($10^{-4} - 10^{-5}$ M) which are adequate for rapid crop growth.

All phosphorus fertilizers contain a proportion of phosphate compounds which are not labile due to their very low solubility products (e.g. hydroxyapatite. Since it is difficult to measure the solubility of phosphates *in soil*, fertilizer manufacturers attempt to predict the availability of fertilizer phosphorus by measuring how much will dissolve in certain aqueous solutions. For example, the available phosphate content of the more soluble fertilizers (superphosphates) is assumed to be the proportion of the phosphate content which is soluble in water, whereas for the less soluble materials (rock phosphate, basic slag) the test solvent is normally a two percent citric acid solution. Typical values for commercial fertilizers are given in Table 18.2.

However, such expressions of availability are not very useful in practice since substances like monocalcium phosphate, $Ca(H_2PO_4)_2$, which are initially available in soils tend to react to give much more insoluble phosphates of iron and aluminium (e.g. Variscite). Consequently, the availability of added phosphate depends upon the period of uptake and the chemical properties of the soil, and is best measured by the amount of phosphorus actually taken up by a crop under standard conditions. Arable crops can recover up to 25% of fertilizer phosphate (superphosphate) in the season of application and much smaller amounts in succeeding seasons.

Because of the low content and availability of phosphorus in rock phosphate and basic slag (Table 18.2), these are of limited use as arable fertilizers, in spite of their low cost. However, the phosphate in these materials is more soluble in acid, than neutral, soils and is released at a show rate over a number of years. Consequently, these materials, as fine powders, are useful in the nutrition of perennial crops, especially upland grassland. All phosphate fertilizers contribute considerable quantities of Ca to the soil, and basic slag is particularly useful in supplying varying amounts of S, Mg and trace elements.

Early investigators thought that P_2O_5 was the active substance in phosphorus fertilizers and so they expressed the total phosphorus content as % P_2O_5. although we now know that phosphate ions are the chemical species taken up by roots, the habit of expressing P content as % P_2O_5 persists to the present. To interconvert % P_2O_5 and %P, we use the following expressions :

$$\% P = \% P_2O_5 \times 0{\cdot}44$$
$$\% P_2O_5 = \% P \times 2{\cdot}27$$

Table 18.3. Properties of potassium fertilizers.

Fertilizer	*Chemical substance*	*Solubility (g 100 ml^{-1})*	*% K*
*Potassium chloride (muriate of potash)	KCl	35	52
Potassium sulphate	K_2SO_4	12	44
Potassium nitrate (saltpetre	KNO_3	13	39

*Obtained from natural deposits of Silvite in Canada, Germany and the former USSR.

POTASSIUM FERTILIZERS

Traditionally, shortage of potassium in soils was alleviated by the application of ashes of wood, crop residues or seaweed, and from this practice comes the name potash. Potassium is now generally supplied in the form of simple, soluble ionic salts, principally potassium chloride, although potassium sulphate or nitrate are used where a crop is sensitive to high chloride levels (e.g. tobacco). In spite of the high solubility of potassium chloride, leaching losses are modest due to the binding of K ions to soil colloids by ion exchange. In the season of application, cereals can absorb 50% and grassland up to 80% of fertilizer potassium from a fertile soil.

As for phosphorus fertilizers, the potassium content of a fertilizer is normally given as the percentage of the oxide, K_2O. To interconvert with % K, the following expressions are used :

$$\% K = \% K_2O \times 0{\cdot}83$$
$$\% K_2O = \% K \times 1{\cdot}2$$

LIMING MATERIALS

Calcium ions must be supplied to agricultural soils for at least three reasons. For :

(a) supply of the macronutrient Ca to plants ;

(b) maintenance of optimum soil pH

(c) maintenance of soil structure (aggregation).

In intensively cultivated soils, these requirements are normally met by regular liming to the required pH. However, in addition to supplying other essential elements (N, P, S etc.), some fertilizers also supply calcium as shown in Table 18.4.

Table 18.4. The calcium content of some fertilizers*.

Fertilizer	*Supplying*	*Mean Ca content (%)*
Nitrochalk	N	8
Gypsum	S	22
Rock phosphate	P	33
Single superphosphate	P	20
Triple superphosphate	P	14
Basic slag	P	32

*Note that these are not liming materials.

Chemical Properties of Liming Materials

Liming materials, which contain calcium ions in association with *non leachable* anions (CO_3, OH), include :

Calcium carbonate, $CaCO_3$, normally as finely crushed limestone, but occasionally as chalk or marl. Dolomitic limestone, a mixed $CaCO_3/MgCO_3$ limestone, has the additional value of supplying Mg. All limestone can be stored and handled without difficulty.

Calcium oxide, CaO (lime, quicklime, burnt lime, unslaked lime) is prepared by the thermal decomposition of limestone in lime kiln :

$$CaCO_3 \rightarrow CaO + CO_2 \uparrow.$$

Quicklime, an unpleasant fine powder which burns the skin, absorbs moisture to give hard lumps of calcium hydroxide and carbonate.

Calcium hydroxide, $Ca(OH)_2$ (slaked lime, builder's lime, hydrated lime), which is prepared by the controlled staking of quicklime ;

$$CaO + H_2O \rightarrow Ca(OH)_2$$

is less caustic to the skin than quicklime.

Although CaO and $Ca(OH)_2$ are more reactive and act more quickly in the soil, ground limestone is the most commonly used agricultural lime due to its lower cost and its superior handling and storage properties.

COMPOUND FERTILIZERS

If two or more nutrients are required by the same crop, then it is normally more convenient and more economic to apply them together as a compound fertilizer. A great variety of different compound fertilizers have been manufactured to meet the requirements of different crops and soils, and their composition is normally expressed by the % N : % P_2O_5 : % K_2O ratio, more commonly written as the NPK ratio.

As a particular example, the ferruginous soils on the plateau areas of Malawi are deficient in phosphorus and nitrogen but rich in potassium, released by the weathering of hornblende minerals. Consequently, the most commonly used general-purpose fertilizer (especially for maize) is 20 : 20 : 0, i.e. 20% N : 20% P_2O_5 : O % K_2O. However, where cotton is grown on hydromorphic soils in the lower altitude areas, it is essential to restrict nitrogen supply to avoid excessive vegetative growth (plants taller than 2 m are not unusual). For this purpose it is customary to use 2 : 18 : 15, i.e. 2% N : 18% P_2O_5 : 15% K_2O.

This ratio gives only the quantities of nutrients present in the fertilizer and reveals nothing about the chemical nature of the ingredients. The 'recipe' for a particular compound fertilizer may vary widely according to :

(a) the *compatibility* of chemicals mixed together, e.g. an acid compound will not be compatible with a basic compound ;

(b) the relative *costs* of different compounds supplying the same element ;

(c) the *commercial availability* of fertilizer compounds.

Most compound fertilizers and many 'straight' (single nutrient) fertilizers are now produced in granular form. Granulation reduces the caking of fertilizer particles, encourages the free running of particles in fertilizer spreaders and cuts down dust hazards. In most compound fertilizers, each granule contains the ratio of nutrients printed on the bag, thus ensuring uniformity of nutrient distribution.